Ultrastructure of Rust Fungi

Ultrastructure of Rust Fungi

LARRY J. LITTLEFIELD
Department of Plant Pathology
North Dakota State University
Fargo, North Dakota

MICHELE C. HEATH
Department of Botany
University of Toronto
Toronto, Ontario, Canada

ACADEMIC PRESS   New York   San Francisco   London   1979
A Subsidiary of Harcourt Brace Jovanovich, Publishers

ACADEMIC PRESS, INC.
111 Fifth Avenue, New York, New York 10003

United Kingdom Edition published by
ACADEMIC PRESS, INC. (LONDON) LTD.
24/28 Oval Road, London NW1 7DX

Library of Congress Cataloging in Publication Data

Littlefield, Larry J
     Ultrastructure of rust fungi.

     Bibliography: p.
     Includes indexes.
     1. Rusts (Fungi) 2. Fungous diseases of
plants. 3. Fungi--Anatomy. 4. Ultrastructure
(Biology) I. Heath, Michele C., joint author.
II. Title.
QK627.A1L57   589'.225   78-22530
ISBN 0-12-452650-0

PRINTED IN THE UNITED STATES OF AMERICA

79 80 81 82   9 8 7 6 5 4 3 2 1

*To the memory of Professor J. J. Christensen,
to Professor R. K. S. Wood, and to all students
of the Uredinales.*

Contents

Preface

Electron microscopy has been employed in the study of rust fungi for some twenty-five years, resulting in a large volume of descriptive and some experimental information. Although several reviews have presented certain features of the ultrastructure of rusts and their associations with higher plants (Beckett *et al.*, 1974; Bracker and Littlefield, 1973; Bushnell, 1972; Coffey, 1975; Ehrlich and Ehrlich, 1971; Hawker, 1965; Petersen, 1974) none has collated the existing information into one source.

The objectives of this book are to provide not only a comprehensive review of rust ultrastructure and host–parasite relations, but also to analyze critically the studies that have been done. We have examined the interpretations and conclusions of investigators based on their own work, on that of other researchers, as well as on our own studies. Sometimes our conclusions do not agree; the reasons for such differences of opinion are given. We have emphasized the dynamics of growth and differentiation rather than just the mature stage of the rusts discussed. We have tried to identify topics in which ultrastructural research is particularly lacking and which provide fertile areas for future research. (Pertinent literature published since preparation of this volume is listed in the Supplemental Bibliography.)

Much of one's understanding of ultrastructural phenomena depends on a concomitant understanding, or at least an awareness, of related physiological and light microscopy research. Where such interrelationships are important to the interpretation of ultrastructural studies we have included discussions of relevant physiological and light microscopy research. Light microscopy is also included where ultrastructural information is scant or absent. This book, however, is not intended as a treatise on cytology or physiology of rust fungi.

We have included considerable unpublished material from our own research where applicable. Numerous tropical species have been described ultrastructurally in an effort to extend the scope of the book beyond the usual temperate rusts with which most North American and European workers are familiar.

We hope this book will provide a stimulus for increased research efforts into the biology of the Uredinales. This work is intended as a reference source for fungal morphologists, taxonomists, plant pathologists, and others interested in the anatomy and associated biology of the rusts.

LARRY J. LITTLEFIELD
MICHELE C. HEATH

Acknowledgments

Excellent technical assistance was provided by Nora Norlan in the preparation of the illustrations and by Susan Willets in the Bibliography. Rosalinda Kloberdanz deserves special recognition for her thorough and conscientious effort in typing the manuscript, aided by Roberta Haspel and Susan Willets.

We are indebted to our many colleagues who contributed micrographs and unpublished information, acknowledged in the text where included. Also, numerous specimens were provided by J. F. Hennen, G. B. Cummins, C. W. Mims, J. D. Miller, and R. W. Stack. R. E. Gold provided excellent assistance in the preparation of many of the previously unpublished scanning electron micrographs. Also, we thank G. B. Cummins, J. F. Hennen, and R. W. Stack for reading portions of the manuscript and for providing constructive criticism. The authors, of course, are solely responsible for any errors or misinterpretations which may appear.

Finally, special thanks go to our spouses: to Julie Littlefield for her understanding patience and to Dr. I. Brent Heath for his support and ideas contributed throughout the preparation of this book.

We acknowledge the following publishers and professional societies that kindly allowed reproduction of micrographs and diagrams from their respective publications: Academic Press, Inc., The Almquist & Wiksell Periodical Company, The American Phytopathological Society, The Botanical Society of America, The Cambridge University Press, Mie University (Japan), The National Research Council of Canada, The Rockefeller University Press, Springer-Verlag, Inc., The Swedish Botanical Society, Verlag Paul Parey, and Wissenschaftliche Verlagsgesellschaft MBH.

LARRY J. LITTLEFIELD
MICHELE C. HEATH

Ultrastructure
of Rust Fungi

1

Introduction

The rust fungi (Class Basidiomycetes, Order Uredinales)* have a complex life cycle that may produce up to five different spore stages,† some of which, depending on the species, form on two different host species obligatorily (heteroecious rusts). Others may pass their entire life cycle on one host species (autoecious rusts). All five spore stages are not present in all rust fungi. A particular nuclear condition, i.e., monokaryotic haploidy, dikaryotic haploidy, or diploidy, is characteristic of each of the spore stages except the teliospores which undergo a transition in nuclear condition during germination. The typical progression of spore stages is basidiospore (N), pycniospore or spermatium (N), aeciospore ($N + N$), urediospore ($N + N$), teliospore ($N + N \rightarrow 2N \rightarrow N$). Species which possess all five spore stages are termed macrocyclic; if the uredial stage is absent, the rust is termed demicyclic; if both the uredial and aecial stages are absent, the rust is designated microcyclic. In some microcyclic rusts, both telia and pycnia are present; in others only telia are present. All microcyclic forms are autoecious, due to their abbreviated life cycle. Macrocyclic and demicyclic rusts may be either autoecious or heteroecious. Examples of the above variations are given in Table 1.

Peterson (1974) and Savile (1976) have provided comprehensive reviews which include variations of the life cycle, the taxonomy, phylogeny, evolution, comparative morphology, and general physiology of the Uredinales.

* Due to the present state of flux of terminology applied to the Division and Class levels of the fungi, a traditional scheme of terminology (Alexopoulos, 1962) is used in this volume.

† Throughout this volume the authors use the system of defining spore stages on the basis of the nuclear cycle, the system established by Arthur (1925) and later expounded by Cummins (1959) and Y. Hiratsuka (1973). Although different terms may be applied to specific stages by the present authors, e.g., pycniospores for spermatia and urediospores for urediniospores, the terms are used in accordance with the basic concepts of the above-mentioned system of nomenclature.

TABLE 1
Variations of the Life Cycle of Rust Fungi

Life cycle	Fungus (example)	Common hosts and spore stages present
Macrocyclic, autoecious	*Melampsora lini*	*Linum* spp. (pycnia, aecia, uredia, telia)
Macrocyclic, heteroecious	*Puccinia coronata*	*Rhamnus* spp. (pycnia, aecia) *Avena* spp. (uredia, telia)
Demicyclic, autoecious	*Gymnoconia peckiana*	*Rubus* spp. (pycnia, aecia, telia)
Demicyclic, heteroecious	*Gymnosporangium cornutum*	*Sorbus* spp. (pycnia, aecia) *Juniperus* spp. (telia)
Microcyclic[a]	*Kunkelia nitens*	*Rubus* spp. (pycnia, telia)
Microcyclic[a]	*Puccinia malvacearum*	*Althaea* spp. (telia only)

[a] All microcyclic rusts are necessarily autoecious.

Those reviews were based on information obtained primarily from light microscopy. The reader is also referred to the classic works by DeBary (1887) and Sappin-Trouffy (1896) which provided the morphological basis upon which modern investigators have had the good fortune to build.

Until the late 1960s, rust fungi were considered to be obligate parasites. Since then numerous species have been grown in axenic culture. The ultrastructure of rusts grown axenically or in tissue culture is essentially the same as those grown parasitically; thus, reference to papers on those topics will be incorporated into the text where appropriate rather than being discussed separately.

The general ultrastructural properties of rusts are similar to other fungi. Likewise, the basic cellular organelles are generally consistent throughout the life cycle of rust fungi. The general ultrastructure of organelles, unless characteristic of particular stages, is discussed in Chapters 4 and 5.

2

Morphology and Ontogeny of Sori and Spores

I. BASIDIOSPORES*

Basidiospores in the rusts are typically uninucleate, haploid spores borne on sterigmata which arise from the linearly arranged four cells of the metabasidium (see Section VI, D of this chapter for discussion of basidium terminology). Exceptions to this include the formation of (a) two binucleate basidiospores per metabasidium, (b) four binucleate basidiospores, (c) two uninucleate and one binucleate basidiospores, or (d) essentially asexual spores in rusts where neither karyogamy nor meiosis occur (Peterson, 1974). Few of the variants have been examined ultrastructurally, other than the binucleate basidiospores of *Gymnosporangium asiaticum* (Kohno *et al.*, 1977b) and *Puccinia horiana* (Kohno *et al.*, 1975a). Other exceptions to the above generality include internal germination of teliospores, e.g., *Coleosporium* spp., where a metabasidium is formed upon septation of the teliospore (probasidium) rather than by growth from the latter (Cummins, 1959). Also, the metabasidium of *Puccinia malvacearum* may fragment in an arthrosporic manner before producing basidiospores (Taubenhaus, 1911). The metabasidium formed by aecioid teliospores of some *Endocronartium* spp. takes the form of a septate germ tube with a single nucleus in each cell (Y. Hiratsuka, 1973).

* Inasmuch as many of the statements about ultrastructural properties of numerous genera and species included in this chapter are previously unpublished, literature references are often not cited. In such cases the reader can correctly assume this to mean "L. J. Littlefield, unpublished."

Basidiospores of *Melampsora lini* are round to oval, approximately 5 × 6 μm, and have a slightly roughened surface and a prominent apiculus or hilar appendage by which the spore is attached to the sterigma [Figs. 1 and 144 (to be discussed in Section IV)].

Basidiospores of *Gymnosporangium juniperi-virginianae* are either pyriform or reniform, generally 10 × 17 μm, and have a prominent apiculus (Mims, 1977a). Although normally uninucleate, those spores may sometimes be binucleate or tetranucleate. Their cell wall consists of a single layer, approximately 0.2 μm thick except around the hilar region where a very thin (ca. 25 nm), electron-opaque layer overlays the relatively electron-lucent inner wall. Three dimensionally the outer wall layer would comprise a cup-shaped covering over the basal end of the basidiospore wall with a circular hiatus in which there is no outer wall layer covering the actual point of attachment of the spore to the sterigma. Such a configuration is consistent with micrographs and the diagrammatic concept of the sterigma–hilum region of *Schizophyllum commune* basidiospores (Wells, 1965). The thin, electron-opaque basal covering may thus represent the remnants of the distal end of the sterigma which previously enclosed the base of the spore rather than a layer of the spore wall per se. No distinct germ pore region exists in *G. juniperi-virginianae,* but germination almost always occurs laterally rather than apically, and the wall of the germ tube is continuous with that of the basidiospore. During germination the large lipid bodies in the cytoplasm are incorporated into vacuoles and subsequently broken down. At that time numerous cytoplasmic vesicles are present in the spore near the germ tube and in the germ tube (Mims, 1977a), presumably playing a role in the apical elongation of the germ tube. Similar observations were reported for germination of *Puccinia horiana* basidiospores except that the binucleate spores typically became tetranucleate during germination (Kohno *et al.,* 1975a). Conversely, the nuclei of binucleate basidiospores fuse to form a single, large nucleus in the germ tube of *Gymnosporangium asiaticum* (Kohno *et al.,* 1977b).

In addition to "direct germination" which produces various types of germ-tube structures, there occurs in some rusts the process of "indirect or repetitive germination." With light microscopy the basidiospores can be observed to produce up to six successive sporidia, each being formed on a small, pointed, sterigmalike appendage of the preceding basidiospore or sporidium, e.g., *Cronartium ribicola* (Bega, 1960). The mode of germination of teliospores of this fungus is influenced by pH, temperature, and the duration of the temperature treatment. Whether this is an important process in nature is not known; however, secondary and tertiary sporidia of *Cronartium fusiforme* are equally pathogenic as the primary basidiospores (Roncadori, 1968).

Upon germination of the basidiospore, the germling produced proceeds commonly to penetrate the host directly through the cuticle and epidermis. An exception to direct penetration is stomatal penetration by basidiospore germ tubes of *Cronartium ribicola* (Fig. 5) (Patton and Johnson, 1970). In *Melampsora lini* a small appressorium is formed at the apex of the germ tube as it penetrates the host epidermal cell wall (Fig. 2). Conversely, the basidiospores of *Gymnosporangium asiaticum* germinate to form an elongated germ tube, at the apex of which is formed a prominent appressorium (Fig. 3) (Kohno *et al.*, 1977b). A similar condition exists in germinated basidiospores of *Cronartium fusiforme* (Fig. 4) (C. G. VanDyke and H. V. Amerson, personal communication). Often, accompanying penetration of the host is an indentation of the host epidermal cell wall and a slight lifting of the fungal structure from the leaf surface shown by light microscopy (R. F. Allen, 1934a,b; Waterhouse, 1921). Waterhouse (1921) was unable to discern any evidence of altered staining properties or cell wall thickness in the region of penetration and concluded that host penetration by basidiospore infection structures is purely a mechanical phenomenon. The progressive increase in resistance of young to mature leaves of *Berberis vulgaris* to basidiospore infection of *Puccinia graminis* f. sp. *tritici* is positively correlated with increased thickness of the outer wall of epidermal cells and increased resistance to mechanical puncture (Melander and Craigie, 1927). However, as discussed by Aist (1976), the significance of this correlation and its original interpretations are questionable. A hypha grows only at its tip, and the resistive force on that hypha remains constant. Conversely, in the case of a needle penetrating a surface, the resistive force increases proportionally to the depth to which the needle penetrates since the resistive force increases according to the surface area of the bore hole. Thus, the observed increase in resistive force may result only from increased thickness in the cell wall and may be unrelated to resistance to infection (Aist, 1976).

II. PYCNIA AND PYCNIOSPORES

Pycnia (= spermagonia) are structures which produce pycniospores (= spermatia), the monokaryotic gametes of rust fungi (Y. Hiratsuka, 1973).

Pycnia commonly begin to develop in herbaceous tissue 4–6 days subsequent to basidiospore infection. In Angiosperms pycnia typically form on the upper leaf surfaces or on fruits. The situation in Gymnosperms is considerably different. There basidiospore infection occurs on the needles, and mycelium may subsequently grow into the woody tissue where pycnia are formed, a process requiring 3–4 years, e.g., *Cronartium ribicola*, where pyc-

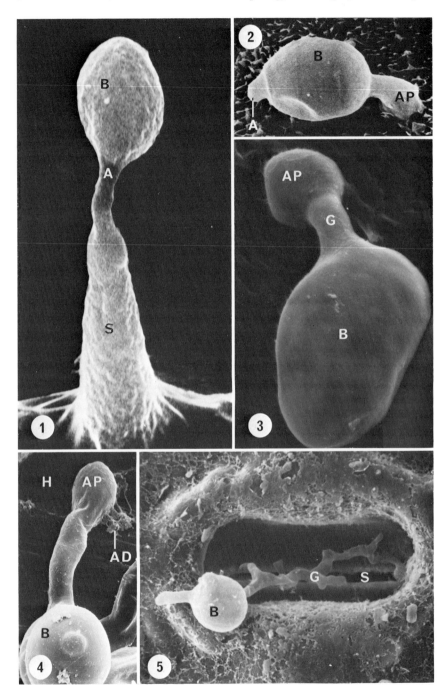

nia develop on branches or the main stem of the tree. In conifer rusts which bear pycnia on the needles, e.g., *Coleosporium* spp., pycnia may form on the current-year needles or on needles infected the previous year (Boyce, 1961).

The morphology of pycnia is variable, depending on the genus and sometimes the species examined. Y. Hiratsuka and Cummins (1963) using light microscopy described 11 types of pycnial morphology, with the growth habit, i.e., determinate or indeterminate, being the character of major importance. Subordinate properties are the presence or absence of bounding structures (e.g., peridium and periphyses), the location of pycnia within host tissue (e.g., subepidermal, subcuticular, intraepidermal, or intracortical), and whether the hymenium is flat or curved into a flask-shaped configuration ("convex" in their terminology). A flask-shaped pycnium of *Puccinia recondita* is shown in Fig. 6. A twelfth type of pycnium characterized by a well developed beak and a large, indeterminate deep-seated cavity (a variant of "Type 4" in the Hiratsuka–Cummins scheme) was recently described with light microscopy (N. Hiratsuka and Y. Hiratsuka, 1977). Numerous pycnia may occur in a pustule which develops from a single-basidiospore infection (Craigie, 1927a). The subglobose pycnium of *Melampsora lini,* with its more or less flat hymenium (Y. Hiratsuka and Cummins, 1963), is shown in Fig. 7.

Various structures differing in morphology and function project through the ostiole of flask-shaped pycnia. Several terms have been applied to those structures, but unfortunately, not in a consistent manner. In *Puccinia* spp., *Gymnosporangium* spp., and some other genera, numerous, pointed, spinelike projections extend radially from the pycnial ostiole. These projec-

---

Fig. 1. Basidiospore (B) of *Melampsora lini* borne at the apex of a sterigma (S), attached by the apiculus (A) at the base of the spore. (× 10,000) [From Gold and Littlefield, 1979. Figures 1 and 2, reproduced by permission of the National Research Council of Canada from *Can. J. Bot.* (in press).]

Fig. 2. Basidiospore (B) of *Melampsora lini* germinates to produce a germ tube with a small, apical appressorium (AP) from which the penetration peg (not visible in this micrograph) originates and penetrates the host epidermis. A, apiculus. (× 4,300) (From Gold and Littlefield, 1979.)

Fig. 3. Germination of basidiospore (B) of *Gymnosporangium asiaticum* to produce a short germ tube (G) and a prominent appressorium (AP). (× 4,000) (From Kohno *et al.*, 1977b.)

Fig. 4. Germination of basidiospore (B) of *Cronartium fusiforme.* Note the apparent adhesive material (AD) in the region of contact between the appressorium (AP) and the host surface (H). (× 2,800) (Courtesy of C. G. Van Dyke and H. V. Anderson, unpublished.)

Fig. 5. Penetration of host by germinating basidiospore (B) of *Cronartium ribicola* through host stoma (S). Note branched germ tube (G) and apparent lack of prominent appressorium compared to Figs. 3 and 4. (× 1,800) (Courtesy of R. F. Patton, unpublished.)

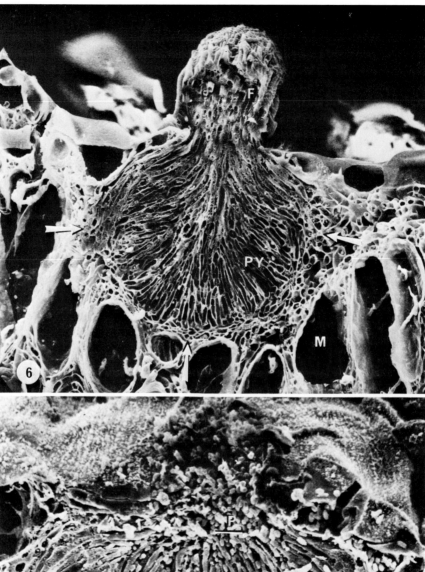

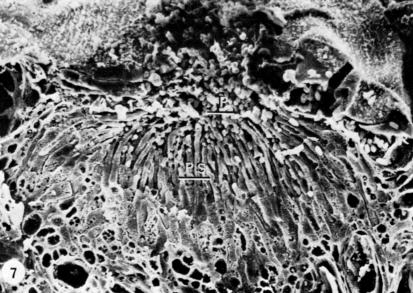

tions extend 75–100 μm beyond the ostiole and may number up to approximately 80 in *P. graminis* f. sp. *tritici* and 140 in *P. minussensis* (Buller, 1950) (see Figs. 11 and 12). A primary function of these projections is to rupture the host epidermis during ontogeny of the pycnium. At an early stage of pycnium development they coalesce to form a rather broad, somewhat pointed aggregation of rigid filaments that provides a most effective structure for penetration of the host epidermis by the expanding pycnium (Fig. 10, *P. recondita*). This aggregation was likened to "a Red Indian tent or tepee" in the descriptive terminology of Buller (1950). It is not until well after epidermal penetration that the projections assume a radiating pattern of distribution (Gold *et al.*, 1979) (Figs. 11 and 12). The ostiolar projections in *Uromyces* spp. and *Puccinia* spp. arise from the sides and roof of the pycnial cavity near the periphery of the ostiole. For these projections Buller used the term "periphyses." In *Gymnosporangium* spp., however, the projections arise from the base of the pycnial hymenium, interspersed among the pycniosporophores, in which case the term "paraphyses" is more accurate. Buller compared the terms, their etymology, and their descriptive accuracy, and concluded that a more general term not requiring histological knowledge for proper usage was required. For this he suggested the terms "ostiolar trichomes" or "ostiolar bristles." Unfortunately he did not follow his own advice, and chose to call all the pointed structures periphyses. Other workers have used all these terms; thus, the literature contains different terms for structures which may or may not be the same, depending on the investigator and the rust fungus being described.

A second type of pycnial projection is flexuous hyphae. In the case of *Melampsora* spp. these are the only pycnial projections present, the periphyses or ostiolar trichomes being absent (Gold and Littlefield, 1979) (Figs. 13 and 14). In most other rusts flexuous hyphae occur intermixed among the ostiolar trichomes (Fig. 12), and are generally less numerous than the ostiolar trichomes. Typically they are sinuous or wavy in outline, hence the designation "flexuous." In *Scopella gentilis*, at least, flexuous hyphae

---

Fig. 6. Cross fracture of a young pycnium (P) of *Puccinia recondita*. A zone of densely packed hyphae (arrows) separates the pycnium from the host mesophyll cells (M). A fascicle (F) of paraphyses and flexuous hyphae extends through the pycnial ostiole and bends onto the leaf surface out of the plane of focus. (× 580) (From Gold *et al.*, 1979. Reproduced by permission of the National Research Council of Canada from *Can J. Bot.* **57**, 74–86.)

Fig. 7. Cross fracture of a young pycnium of *Melampsora lini*, bearing pycniospores (P) borne on elongated pycniosporophores (PS). Note the comparatively flat nature of this pycnium contrasted to the typically globose pycnium of *Puccinia recondita* as shown in Fig. 6. (× 860) [From Gold and Littlefield, 1979. Reproduced by permission of the National Research Council of Canada from *Can. J. Bot.* (in press).]

arise from modified pycniosporophores in rather mature pycnia, as shown by light microscopy (Payak, 1956).

Flexuous hyphae classically function as the female structures onto which the male pycniospores (spermatia) attach and transfer their nucleus during plasmogamy. However, in *Melampsora* spp. the flexuous hyphae may have no reproductive function as plasmogamy is claimed to be effected by another mechanism, according to R. F. Allen (1934a), although evidence is lacking to substantiate this (see Section III, this chapter). Also, the flexuous hyphae of *Melampsora lini* are not involved in penetrating the host epidermis since host stomata serve the function of pycnial ostioles (Figs. 13 and 14).

There may be up to 50 flexuous hyphae per pycnium in *Melampsora lini*; they may be branched or unbranched, their diameter is approximately 1.0–1.5 μm, and they vary in length up to approximately 60 μm (Fig. 14). The number of flexuous hyphae per pycnium is less in *Gymnosporangium clavipes* but they have similar morphology (Kozar and Netolitzky, 1975). From light microscopy, Buller (1950) reported less than 10 flexuous hyphae per pycnium for *Phragmidium speciosum*, *Gymnoconia peckiana*, and *Tranzschelia pruni-spinosae*, but as many as 20 in *Puccinia graminis*.

The properties of flexuous hyphae compared to ostiolar trichomes (periphyses and paraphyses), visible at the light microscope level, were summarized by Buller (1950) as follows

. . . (1) they are flexuous instead of being straight or slightly curved; (2) they are cylindrical and have bluntly rounded ends instead of being slenderly conical and tapering to a point; (3) they grow in length for some time and become longer than the periphyses; (4) occasionally they are branched, whereas periphyses are never branched; (5) they are less red than periphyses; (6) they fuse with pycni(di)ospores of opposite sex, whereas periphyses do not; and (7) they are relatively ephemeral, so that in old pycn(id)ia of Puccinia, etc., one can find periphyses but no flexuous hyphae.

Fig. 8. Pycniosporophores (PS) of *Melampsora lini* are often branched. Some pycniospores are immature and still attached to the sporophores (open arrows); mature pycniospores (solid arrow) are free in the pycnial cavity. ($\times$ 1,800) [From Gold and Littlefield, 1979. Reproduced by permission of the National Research Council of Canada from *Can. J. Bot.* (in press).]

Fig. 9. Mature pycniospores (P) of *Puccinia recondita* on the host surface. They are typically smooth-surfaced and oval in shape. ($\times$ 3,750) (From Gold *et al.*, 1979. Figures 9 and 10, reproduced by permission of the National Research Council of Canada from *Can. J. Bot.* 57, 74–86.)

Fig. 10. A young pycnium of *Puccinia recondita* as seen on the host surface. The column consists of paraphyses and flexuous hyphae united in a fascicle with a honeydew condensate (H). Apically, the hyphalike components extend beyond the surface of the condensate. ($\times$ 900) (From Gold *et al.* 1979.)

Fig. 11. A slightly later stage of pycnial development in *Puccinia graminis* f. sp. *tritici* compared to Fig. 10. The paraphyses (PH) have begun to separate, and the honeydew condensate (H) occurs only near the apex of the fascicle. ($\times$ 520) (L. J. Littlefield, unpublished.)

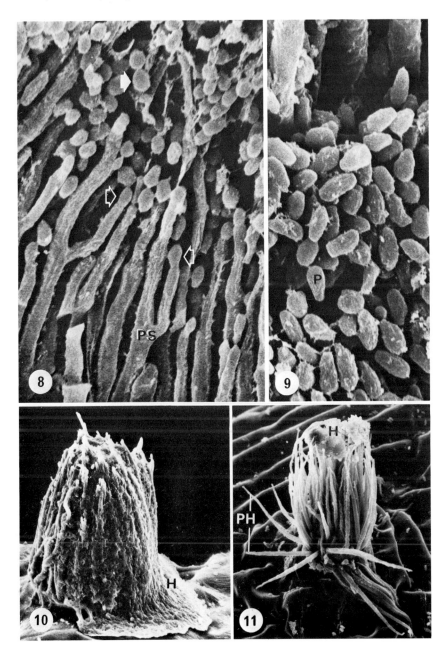

An exception to characteristic 3 exists with *Puccinia recondita,* where the flexuous hyphae may be shorter or longer than the ostiolar trichomes (Fig. 12) (Gold *et al.,* 1979).

Mature ostiolar trichomes in older pycnia may be transformed into flexuous hyphae (Buller, 1950); however, this has yet to be confirmed by electron microscopy. This conversion is effected by renewed growth of ostiolar trichomes at their apices, but in the inflated, cylindrical, and wavy configuration of flexuous hyphae.

Although ostiolar trichomes (periphyses and paraphyses) may be absent in some species, flexuous hyphae apparently are present in most if not all Uredinales. Buller (1950) and Payak (1956) described and illustrated these structures using light microscopy on 53 species in 16 genera in the Melampsoraceae and Pucciniaceae.

Pycniospores function as monokaryotic gametes which are produced at the tips of pycniosporophores lining the hymenial surface of the pycnium (Figs. 8, 15-18). Pycniosporophores are often long cells (18-33 μm, calculated from published micrographs and drawings) which taper slightly distally (Figs. 8 and 15). In *Puccinia coronata* f. sp. *avenae* they arise from a layer of irregularly shaped pseudoparenchymatous cells, with many cells of the latter dividing to form several pycniosporophores (Harder and Chong, 1978). Pycniospores are produced in basipetal succession (Figs. 8 and 15) from the tips of the sporophores, each one being uninucleate following the mitotic division of the nucleus in the sporophore prior to spore cleavage.

Immediately prior to formation of the first pycniospore the nucleus in the pycniosporophore undergoes a single mitotic division and the apex of the sporophore becomes swollen to form the pycniospore initial. One daughter nucleus migrates into the spore initial and the other remains in the sporophore (Fig. 16). A septum is then formed at the base of the spore initial

---

Fig. 12. Surface view of a mature pycnium of *Puccinia recondita.* The pointed paraphyses (PH) and the sometimes branched flexuous hyphae (F) radiate from the ostiole (O) which was formed upon rupture of the host epidermis (E). (× 860) (From Gold *et al.,* 1979. Reproduced by permission of the National Research Council of Canada from *Can. J. Bot.* 57, 74-86.)

Fig. 13. An early stage in the extension of flexuous hyphae (F) of *Melampsora lini* through a stoma on the host leaf. Note the enlarged apices of such immature flexuous hyphae compared to their filamentous nature at maturity, Fig. 14. (× 1,150) [From Gold and Littlefield, 1979. Figures 13 and 14, reproduced by permission of the National Research Council of Canada from *Can. J. Bot.* (in press).]

Fig. 14. Surface view of a mature pycnium of *Melampsora lini.* The slender flexuous hyphae (F) extend through a leaf stoma and are surrounded by numerous pycniospores (P). The honeydew matrix was removed during specimen preparation. (× 1,070) (From Gold and Littlefield, 1979.)

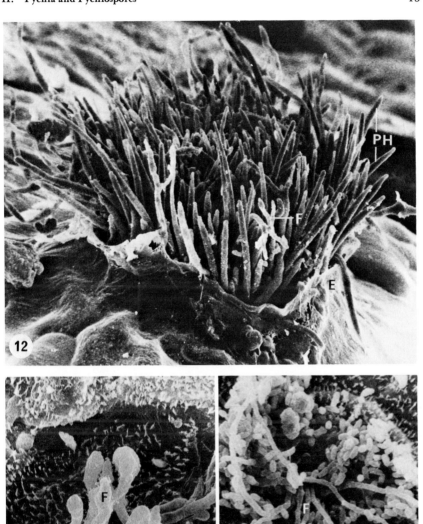

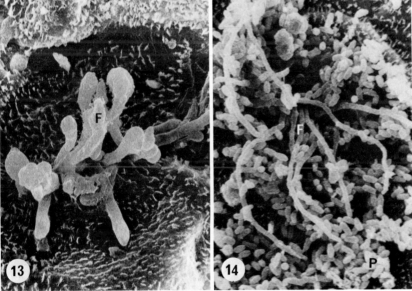

which separates the immature pycniospore from the pycniosporophore. The septum is complete, i.e., without a central perforation; it is formed by localized centripetal invagination of the cell wall, and consists of two wall layers separated by an electron-lucent lamella which extends outward to contact the inner face of the outer layer of the periclinal wall. Disjunction of the immature spore from the sporophore is effected by cleavage along the central lamella of the septum which separates the two structures. Ultimately this cleavage ruptures the outer cell wall layer, leaving a pronounced basal frill on the wall around the base of the pycniospore. The basal frill consists of remnants of the ruptured outer cell wall layer. The spores are thereby released into the pycnial cavity where they undergo further maturation, marked by the thickening of the inner, secondary cell wall layer and the diminution of the outer, primary cell wall layer, including the basal frill. During these simultaneous changes the cell wall does not change drastically in thickness, remaining about 45 nm thick, but the inner wall layer replaces the outer wall layer as the one comprising the bulk of the spore wall (Harder and Chong, 1978).

The pycniosporophores continue to produce successional pycniospores. The successive cleavages of spores leave pronounced rings, or annular scars, at the sporophore apex (Figs. 16 and 17), all borne at approximately the same level. As many as 9–11 such annular scars have been observed in pycniosporophores of *Puccinia coronata* f. sp. *avenae* (Harder and Chong, 1978). The formation of the initial and successive pycniospores is diagrammed in Fig. 18.

Rijkenberg and Truter (1973b, 1974a) proposed that pycniosporogenesis in *Puccinia sorghi* and *Gymnosporangium* spp. (the latter based on reinter-

---

Figs. 15–17.   Pycniosporogenesis in *Puccinia coronata* f. sp. *avenae*. (From Harder and Chong, 1978. Reproduced by permission of the National Research Council of Canada from *Can. J. Bot.* **56**, 395–403.)

Fig. 15.   Longitudinal section through a pycniosporophore (SP₁) that has given rise to a pycniospore initial (PI₁). An adjacent pycniosporophore (SP₂) bears a pycniospore initial (PI₂) and a chain of pycniospores (P). Note the flared collar (arrows) at the apices of the pycniosporophores from which the spore initials have arisen. (× 6,000)

Fig. 16.   A pycniospore initial (PI) at the apex of a pycniosporophore (SP). The nucleus (N) has partially migrated into the pycniospore initial. The cell wall of the latter is continuous with the innermost layer of the flared, multilayered collar (arrow) at the apex of the sporophore. (× 13,800)

Fig. 17.   The multilayered collar (C) at the apex of the pycniosporophore (SP) consists of a series of annular scars left by a succession of pycniospore initials. The wall of each succeeding pycniospore initial (PI) is continuous with the innermost wall layer of the collar (arrow). (× 33,800)

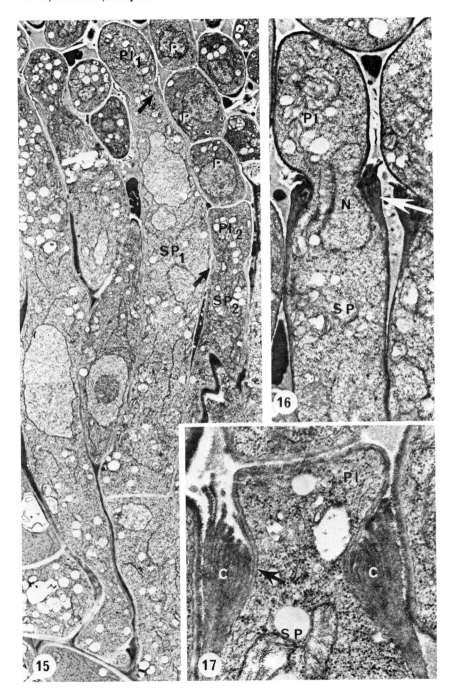

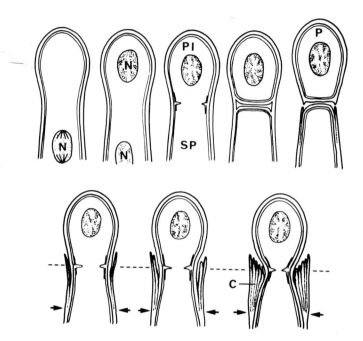

Fig. 18. A diagrammatic summarization of pycniospore formation in *Puccinia coronata* f. sp. *avenae.* The upper row illustrates stages in the formation of the first pycniospore, and the lower row shows the wall relationships during the formation of successive pycniospores. The dotted line in the lower row shows the level at which septation occurs, and the arrows indicate the approximate point of origin of the walls of all succeeding pycniospores. C, collar; N, nucleus; P, pycniospore; PI, pycniospore initial; SP, pycniosporophore. (From Harder and Chong, 1978. Reproduced by permission of the National Research Council of Canada from *Can. J. Bot.* **56,** 395–403.)

pretation of early light microscopy) is annellidic. The same conclusion was drawn by Harder and Chong (1978) regarding *Puccinia coronata* f. sp. *avenae,* although some phialidic ontogenetic properties are exhibited in that fungus.

The conceptual delineation between phialosporogenesis and annelosporogenesis is not always clear. In some instances a clear distinction may be possible, but as Hammill (1974) suggests, there may exist a continuum between these types of development, with some polymorphic fungi occupying different positions on the continuum, depending on environment, age, or other factors. In the few rust fungi studied to date, the first pycniospore to be formed from the sporophore is clearly annellidic, based on its holoblastic nature (all cell wall layers of the spore being derived entirely from the parent sporophore wall). However, the successive spores, also annellidic, arise from essentially the same level of the sporophore apex, the latter not having elongated with production of each successive spore, a characteristic of phialidic development.

Upon cleavage of pycniospores from pycniosporophores the former are released into the pycnial cavity in a matrix of honeydew which fills the cavity. The honeydew and the pycniospores exude through the pycnial ostiole onto the surface of the host plant (Figs. 9 and 14), although a discrete ostiole may be absent in some types of pycnia (Y. Hiratsuka and Cummins, 1963).

III. THE DIKARYOTIZATION PROCESS

Unfortunately, knowledge of the dikaryotization process is derived almost entirely from light microscopic studies, which ceased essentially in the 1930s. It is only necessary to describe the process briefly since it has been reviewed thoroughly by Lamb (1935) and Buller (1950) and little new information has been added since 1950, except for two studies (Craigie, 1959; Craigie and Green, 1962). Although electron microscopy has not been employed, the inclusion of such information obtained by light microscopy is justified in a book on rust ultrastructure since dikaryotization is an essential feature of the complete life cycle of rust fungi.

Although Craigie (1927a,b) proved with *Puccinia helianthi* and *P. graminis* f. sp. *tritici* the sexual function of pycnia, the actual fusion of a pycniospore with a flexuous hypha was not shown until 6 years later (R. F. Allen, 1933; Craigie, 1933; Pierson, 1933). Fusion in *Puccinia helianthi* was by means of a small "fusion tube" between the pycniospore and the flexuous hypha, approximately 2.0 μm long, calculated from Craigie's light micrograph. A similar connecting structure was illustrated in some, but not all, the fusions shown by R. F. Allen (1935b) with *Puccinia sorghi.* Pierson (1933) described a similar phenomenon in *Cronartium ribicola,* although the fusion tubes described were longer and somewhat narrower in diameter than in *Puccinia helianthi.* Savile (1939) illustrated attachment of pycniospores and flexuous hyphae accompanied by the flow of the nucleus and cytoplasm from the former into the latter. There were no fusion tubes reported, only slightly raised, circular papillae on the wall of the flexuous hypha at the points of pycniospore attachment. A more comprehensive study of this topic was made by Buller (1950) with *Puccinia graminis* f. sp. *tritici, P. coronata* f. sp. *avenae, P. helianthi,* and *Gymnosporangium clavipes.* Except for *P. helianthi,* where fusion nearly always occurs at the tips of the flexuous hyphae, the pycniospores may fuse with flexuous hyphae anywhere along their length. Apparently, according to Buller, a pycniospore in close proximity to a flexuous hypha sends out a stimulus that induces the latter to initiate a branch which grows toward the pycniospore, or causes the bending of the hyphal tip to the same. Likewise, the flexuous hypha sends a stimulus which induces the pycniospore to form a tiny, blunt papilla which meets with and fuses with the oncoming hyphal tip. Buller (1950) was unable to induce pycniospores to germinate and form germ tubes, therefore, concluding that the fusion tube is presumably produced entirely by the flexuous hypha except for the small papilla on the pycniospore wall.

Generally, only one pycniospore may fuse with each flexuous hypha (Buller, 1950), although a double fusion was suggested in some instances by Savile (1939). Multiple fusions can occur within the population of flexuous hyphae in a pycnium, however, resulting in various genotypes of aeciospores per aecium (Newton *et al.*, 1930).

What appears to be the fusion of a pycniospore and a flexuous hypha of *Melampsora lini* is shown in Fig. 19. An apparent fusion tube, approximately 1.8 μm long, joins the two structures. Only one such observation has been made in our studies (R. E. Gold and L. J. Littlefield, unpublished); repeated observations by transmission and scanning electron microscopy are necessary to confirm this tentative conclusion.

In the interim between Craigie's discovery of the function of pycniospores (1927a,b) and the demonstration of fusion between pycniospores and flexuous hyphae (R. F. Allen, 1933b; Craigie, 1933; Pierson, 1933), numerous proposals were made to explain the mechanism of dikaryotization in rust fungi. Germination of pycniospores to form narrow hyphae which grow into

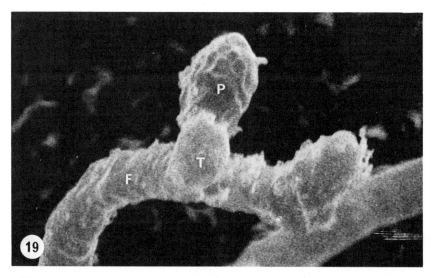

Fig. 19. The apparent fusion of a pycniospore (P) and a flexuous hypha (F) of *Melampsora lini*, connected by a fusion tube (T). (× 11,000) (R. E. Gold and L. J. Littlefield, unpublished.)

the host intercellular spaces and there fuse with haploid mycelium of the opposite mating type was proposed for *Puccinia graminis* f. sp. *tritici* (R. F. Allen, 1933a; Hanna, 1929) and *Melampsora lini* (R. F. Allen, 1934a). The fusion of pycniospores with so-called "receptive hyphae" apart from pycnia was proposed by R. F. Allen (1932a,b, 1933a, 1934b) for *Puccinia recondita, P. coronata* f. sp. *avenae, P. graminis* f. sp. *tritici,* and *P. sorghi,* respectively, and for *Uromyces* spp. by Andrus (1931, 1933). The above-mentioned mechanisms proposed by R. F. Allen (1932a,b, 1933a, 1934a,b), Andrus, (1931, 1933), and Hanna (1929) have not been corroborated by subsequent investigators.

After thorough examination of the literature, coupled with his own experiments, Buller (1950) accepted the following modes of initiating the dikaryon as having been sufficiently documented: (I) fusion of (+) or (−) pycniospores, respectively, e.g., *Puccinia helianthi* (Brown, 1935), *P. helianthi,* and *P. graminis* f. sp. *tritici* (Craigie, 1927a,b); (II) fusion of (+) or (−) pycniospores, respectively, with (−) or (+) flexuous hyphae; e.g., *Puccinia graminis* f. sp. *tritici* (Craigie, 1927a,b 1933); (III) fusion of $N + N$ mycelium, derived from an aeciospore or a urediospore, with a haploid mycelium derived from a basidiospore; restricted to autoecious rusts; e.g., *Puccinia helianthi* (Brown, 1932); (IV) pairing of nuclei within haploid mycelium of microcyclic rusts prior to the formation of teliospores, e.g., *Puccinia malvacearum* (R. F. Allen, 1935), or by the pairing of nuclei in the basidium to produce only two, binucleate basidiospores per basidium, e.g., *Puccinia arenariae* (Lindfors, 1924, cited by Buller, 1950); (V) fusion by modes I, II or III, but following the annual de-dikaryotization (dediploidization, Buller's terminology) of certain rusts having a systemic mycelium; e.g., *Puccinia minussensis* (Brown, 1941b).

To this list must be added the fusion of germ tubes from aeciospores and urediospores with flexuous hyphae of *Puccinia graminis* f. sp. *tritici,* demonstrated by Cotter (1960) and confirmed by Garrett and Wilcoxson (1960).

It is essential to keep an open mind on the subject of mechanisms by which dikaryotization occurs. Granted, convincing evidence is lacking to support R. F. Allen's suggestion of pycniospore germination, host penetration, and subsequent fusion with a compatible hypha. Buller (1950) severely criticized the idea of extrapycnial "receptive hyphae" as sites for spermatization (R. F. Allen, 1932a,b, 1933a, 1934b; Andrus, 1931, 1933). However, adequate evidence to disprove the validity of these observations and their resultant conclusions is likewise lacking. The full range of dikaryotization mechanisms in the Uredinales is probably yet to be discovered.

What happens, precisely, to the pycniospore nucleus, once it enters the "female" thallus is not entirely clear, although light microscopy by Craigie

(1959) and Craigie and Green (1962) have provided considerable insight into that question. As those authors recognized, the unequivocal identification of pycniospore-derived nuclei versus resident nuclei in the haploid mycelium is most difficult, although the "unexpanded" nature of the former versus the "expanded" nature (sensu Savile, 1939) of the latter provide a reasonably reliable criterion. In both *Puccinia helianthi* (Craigie, 1959) and *P. graminis* f. sp. *tritici* (Craigie and Green, 1962) pycniospore nuclei, subsequent to the fusion of pycniospores and flexuous hyphae, migrate through the haploid hyphae into the protoaecium. This migration requires some 20–25 hours and is accompanied by little mitotic activity. The intercellular hyphae remain haploid, with only transitory dikaryons being formed during the migration process. Once the pycniospore nuclei reach the cells at the base of the pro-toaecium, stable dikaryons are established. This is followed rapidly by numerous, consecutive, conjugate nuclear divisions and subsequent migra-tions of daughter nuclei into other cells of the protoaecium. This provides numerous multinucleate cells from which the dikaryotic aeciospore initials are derived. A great void exists, however, in a detailed knowledge and understanding of these processes. Critical ultrastructural research is required to answer the many yet unresolved questions concerning dikaryotization in the rust fungi some half century after Craigie's (1927a,b) discovery of the sex-ual function of pycnia and pycniospores. Perhaps methods by which nuclei from pycniospores can be differentially labeled, e.g., microautoradiography or fluorescent antibody techniques, could be used to study the transfer of nuclei between sexually compatible thalli and the fate of such nuclei once in the "female" thallus. Accompanying such studies, of course, must be thorough documentation of the hyphal connections which link pycnia and aecial primordia.

IV. AECIA AND AECIOSPORES

Aecia are sori which produce nonrepeating, vegetative aeciospores, usually as a result of dikaryotization, and are thus typically associated with pycnia (Y. Hiratsuka, 1973). In rusts which produce the aecial stage, the aecia often form within a few days following the appearance of pycnia. Typically the aecia form on the lower surface of leaves beneath the pycnia on the upper surface, although it is not uncommon for aecia to form on the upper surface of leaves around the pycnia. Aecia may also form on stems of herbaceous Angiosperms and woody Gymnosperms. In coniferous hosts, aecia often oc-cur in association with localized swellings or galls, e.g., *Cronartium fusiforme* and *C. cerebrum,* or necrotic cankers, e.g., *Cronartium ribicola.* Aecia may even form on the internal surfaces of mummified fruits and on the seeds of *Ribes glandulosum* infected with *Puccinia albiperdia* (Taylor, 1921). *Puccinia angustata* forms internal aecia within the host stem and petiole pith as well as on the outer surfaces of those organs (Wolf, 1913). Numerous aecial primordia may form within a single sorus following inoculation with a single basidiospore (Buller, 1950), although the aeciospores produced in different aecia of a single sorus may be different genetically due to fertilization of the associated pycnia by genetically different pycniospores (Newton *et al.*, 1930).

In rusts which produce aecia on conifer stems the aecia typically do not appear until the season following the production of pycniospores, the latter not occurring until 3–4 years after basidiospore infection (Boyce, 1961).

A. Aecial Types

Aecia vary in their overall morphology. They can be grouped into five major morphological types, depending primarily on the absence or presence of a peridium, and the nature of the peridium if present (Cummins, 1959).

1. Caeomoid Aecia

The caeoma, or "caeomoid aecium," is the least organized type, having an irregular outline due to the lack of a well defined peridium. Although *Melampsora* spp. caeoma (Figs. 20 and 21) are defined as nonperidiate (Cummins, 1959), they may have in some cases a rudimentary peridium (Wilson and Henderson, 1966) consisting of a single cell layer that is visible in the light microscope (Hersperger, 1928).

2. Aecidioid Aecia

The "aecidioid aecium," the stereotypic form of aecia commonly illustrated in textbooks, is typical of *Puccinia* spp. and *Uromyces* spp. Many individual aecia may be aggregated into one aecial sorus. The aecidioid aecium is a cylindrical to cup-shaped structure with aeciospores borne within the confines of the peridium, a layer one cell thick, which clearly delimits the peripheral boundaries of the aecium (Figs. 22–24).

The peridial cells are usually rhomboidal, and their cell walls are differentially thickened. The wall on the inner face, i.e., the side toward the aeciospore columns, is approximately 1 μm thick, but measures up to 8 μm thick on the outer face, e.g., *Puccinia sorghi* (Rijkenberg and Truter, 1974b) and *P. recondita* (Fig. 33) (Gold *et al.*, 1979). The thick, inner-wall face consists of an electron-opaque wall matrix containing extensive, dagger-shaped, electron-lucent processes which extend radially, but not beyond the outer surface of the cell wall (Fig. 33). On the thin, opposite wall, in contrast, the electron-opaque matrix is only approximately 1 μm thick and the irregular, electron-lucent, knoblike processes extend approximately 1–2 μm beyond the

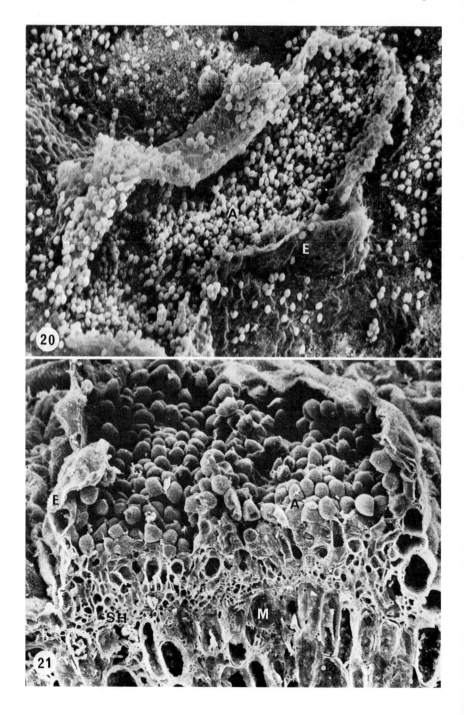

wall surface to provide the surface ornamentation of the peridial cells (Figs. 32-35). The peridial cell ornaments are larger and less delicate compared to the ornaments of the neighboring aeciospores (Fig. 32), although their ontogeny shares several common features (Gold *et al.*, 1979); see Section IV,C, this chapter.

The length to which the aggregated aeciospore columns and the surrounding peridium *Puccinia graminis* f. sp. *tritici* aecia extend depends on two major factors, length of time after dikaryotization and the ambient humidity (J. D. Miller and L. J. Littlefield, unpublished). If the infected host is kept in relatively low humidity the peridium may remain intact for several days, reaching a length of some 3-4 mm (Fig. 23). However, if such aecia are transferred to a highly humid environment the peridium soon ruptures and aeciospores are forcibly discharged. If the aecia are kept at high humidity from the outset they rupture soon after emergence from the leaf and never attain the lengths possible when maintained at low humidity.

3. Roestelioid Aecia

The "roestelioid aecium," characteristic of *Gymnosporangium* spp., is a hornlike structure, often 1-3 mm long, surrounded by a tubular, one-cell-thick peridium (Kunoh *et al.*, 1977). The peridium may cleave longitudinally into fine, recurved, hairlike structures, e.g., *Gymnosporangium juniperavirginianae* (Figs. 25 and 27), or it may simply disintegrate into small, longitudinal shreds from the apex downward, without recurving, e.g., *Gymnosporangium globosum* (Figs. 26 and 28). Exceptions to this typical longitudinal disintegration of peridial cells are found in *Gymnosporangium cornutum* and *Gymnosporangium fuscum* (Parmelee, 1965; Leppik, 1977; Ziller, 1961), shown by light microscopy. In the former the cornute aecia retain their cylindrical shape by rupturing along the sides, and aecia of the latter shred longitudinally similar to most Gymnosporangia, but the apex remains joined, resulting in the characteristic balanoid (acorn-shaped) configuration.

The structure of peridial cells of roestelioid aecia is employed as a useful taxonomic character in light microscopy (Kern, 1910, 1973; Parmelee, 1965). Although sometimes rhomboidal, the peridial cells of rostelioid aecia

---

Fig. 20. Surface view of a caeomoid aecium of *Melampsora lini*. The host epidermis (E) is ruptured, revealing the aeciospores (A) within the sorus. × 1,300. [From Gold and Littlefield, 1979. Figures 20 and 21 reproduced by permission of the National Research Council of Canada from *Can. J. Bot.* (in press).]

Fig. 21. Cross fracture of an aecium of *Melampsora lini* in approximately the same stage of maturity as that in Fig. 20. Sporogenous hyphae (SH) at the base of the sorus give rise to short chains (seldom more than 2-3 cells) of aeciospores (A). E, ruptured host epidermis; M, host mesophyll cells. (× 350) (From Gold and Littlefield, 1979.)

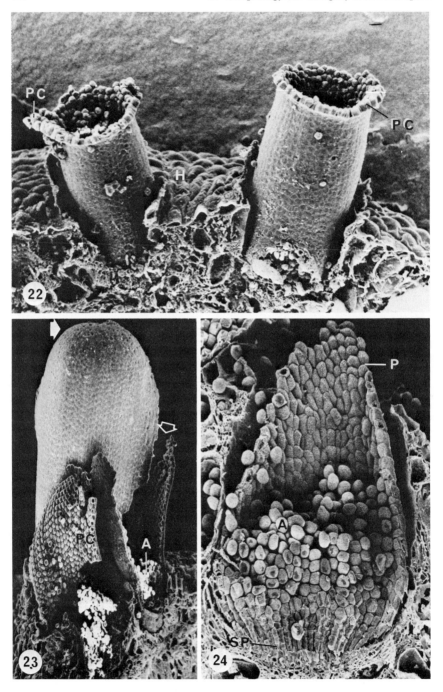

are typically elongated, 80–300 μm, and arranged end to end, forming long files of cells which separate longitudinally to produce the characteristic lacerate or fimbriate appearance (Fig. 25). The ends of the cells are over-lapped (imbricated) (Fig. 27), giving an articulated nature that provides for movable joints, allowing the ruptured peridium to twist and curl (Kern, 1973). The walls of the peridial cells are greatly thickened on the two lateral sides and the inner face toward the aeciospores. These are also the three sur-faces of the peridial cells which are usually ornamented. The ornamentation may be verrucose, tuberculate, papillose, or ridged (rugose), or in rare cases, the surface may be smooth (Kern, 1910). Kozar and Netolitzky (1975) il-lustrated clavate projections approximately 1.5 μm tall on the surface of peridial cells of *Gymnosporangium juniperi-virginianae,* similar to those of *Puccinia recondita* (Fig. 32) (Gold *et al.,* 1979).

4. Peridermioid Aecia

The gross morphology of the "peridermioid aecium" commonly resembles a tongue or a blister flattened in one plane (Figs. 29–31). They are typical of *Cronartium* spp., *Coleosporium* spp., *Milesia* spp., and *Pucciniastrum* spp. (Cummins, 1959). The peridium consists of several layers of thick-walled cells which cover the entire surface of the aecium which led to the early designa-tion of such aecia as "operculate" (Arthur, 1929). What could be interpreted as an "operculum" is often observed in *Cronartium ribicola* when a portion of the disintegrated peridium remains partially attached to the aecium (Fig. 31). The peridium fragments at maturity (Figs. 30 and 31), thus allowing dispersal of the enclosed aeciospores. The peridium forms by the differentia-tion of the distal-most cells of the aeciospore chain into the typical thick-walled peridial cells (Colley, 1918), similar to peridium formation in uredia of *Melampsora* spp. (see Section V, this chapter). This contrasts clearly with the mode of peridium morphogenesis in aecidioid and rostelioid aecia where

---

Fig. 22. Aecidioid aecia of *Puccinia graminis* f. sp. *tritici* extended as columnar structures several millimeters beyond host surface (H). The outer limits of the aecia consist of tightly bound peridial cells (PC) which form the peridium. (× 135) (L. J. Littlefield, unpublished.)

Fig. 23. Prior to rupture of the peridium, the aecium of *Puccinia graminis* f. sp. *tritici* is a cylindrical structure, intact at the apex (solid arrow). The aecium in the rear has ruptured longitudinally (open arrow) during preparation, revealing the aeciospores (A) enclosed therein. Only remnants of the aecium in front (also damaged during preparation) remain, revealing the closely packed peridial cells (PC) which line the inside of the structure. (× 90) (L. J. Littlefield, unpublished.)

Fig. 24. Longitudinal fracture of an aecidioid aecium of *Puccinia recondita.* Long chains of aeciospores (A) are produced from aeciosporophores (SP) at the base of the sorus. P, peridium comprised of closely spaced peridial cells. (× 300) (From Gold *et al.,* 1979. Reproduced by per-mission of the National Research Council of Canada from *Can. J. Bot.* **57,** 74–86.)

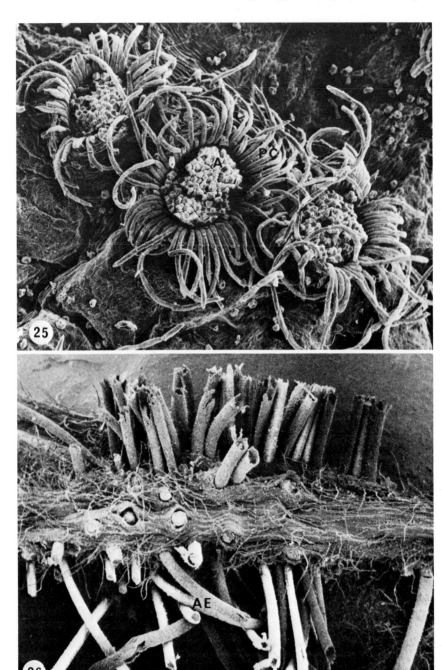

peripheral chains of aeciospores and intercalary cells differentiate to form the tubular peridium around the elongated aecia (Fromme, 1914).

5.   Uredinoid Aecia

A fifth type of aecium, the "uredinoid aecium," lacks a peridium and bears pedicellate, echinulate, nonrepeating aeciospores that appear morphologically similar to most urediospores. They are typical of *Prospodium* spp., *Uropyxis* spp., *Trachyspora* spp., *Pileolaria* spp., and other genera (Cummins, 1959), and are indistinguishable morphologically from uredia. Other than one study dealing in part with ornamentation of the uredinoid aeciospores (see Section C, this Chapter) of *Trachyspora intrusa* (Henderson, 1973), uredinoid aecia have not been examined ultrastructurally.

B.   Aeciospore Formation

The most thorough ultrastructural documentations of aeciosporogenesis are those for *Puccinia sorghi* (Rijkenberg and Truter, 1974b, 1975) and *Puccinia recondita* (Gold *et al.*, 1979).

In *Puccinia sorghi* the base of the aecial stroma consists of a mass of interwoven plectenchymous cells of four main types, (a) uninucleate, often vacuolate cells, (b) multinucleate fusion cells, (c) binucleate aeciosporophores, and (d) cells with degenerate cytoplasm and large vacuoles containing a granular residue (Fig. 36). Cell walls between adjacent uninucleate cells undergo lysis, resulting in the formation of the large multinucleate fusion cells (Fig. 37) which have commonly been observed with the light microscope, e.g., Colley (1918) and Fromme (1914). The fusion cells give rise to the binucleate primary aeciosporophores (Figs. 36 and 38). Exactly how the several nuclei of the fusion cells separate out to form binucleate sporophores is not known. Some such mechanism must exist, however, to provide for the isolation of compatible mating-type nuclei into the sporophores since they ultimately give rise to dikaryotic aeciospores, each of which carries two nuclei of different mating types. The two nuclei in the primary aeciosporophore undergo conjugate division, with one pair of

---

Fig. 25.   Roestelioid aecia of *Gymnosporangium juniperi-virginianae*. Aeciospores (A) are borne at the base of the tubular sori. The peridium consists of parallel files of elongated peridial cells (PC) which recurve in low humidity as shown here, but uncurl to make more or less continuous, tubular structures during periods of high humidity. See Fig. 27 for details of peridial cell ornamentation and their imbricate relationship with one another. (× 100) (L. J. Litlefield, unpublished.)

Fig. 26.   Roestelioid aecia (AE) of *Gymnosporangium globosum*. Aeciospores contained within these sori are released as the apices of the aecia progressively disintegrate; see Fig. 28. (× 20) (L. J. Littlefield, unpublished.)

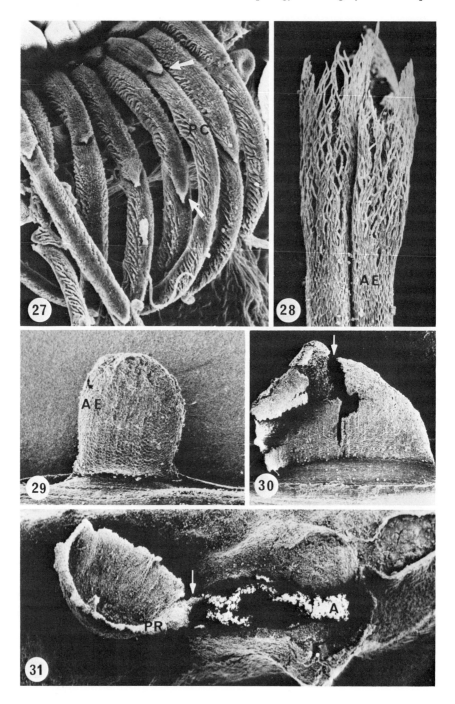

daughter nuclei migrating to the distal end of the cell and one pair remaining in the proximal end of the cell. The primary aeciosporophores may branch basally, proliferating to form numerous binucleate secondary aeciosporophores. Following conjugate nuclear division in the aeciosporophore, the aeciospore initial is separated from the sporophore by a newly formed septum which compartmentalizes the two distal nuclei into the aeciospore initial and the two proximal nuclei into the aeciosporophore (Fig. 38). The septum separating the aeciospore from the aeciospore initial cleaves and separates the former from the latter. Remnants of the periclinal walls in the region of cleavage form annulations on the aeciosporophore wall, thus leading Rijkenberg and Truter (1973b, 1974b) to conclude that aeciospore *initials* (not aeciospores) are annellophoric in their mode of development, not meristem arthrosporic as described by Hughes (1970) with the light microscope. With each successive separation of an aeciospore initial from the aeciosporophore, an annulus remains on the wall of the latter, forming a thick-walled collar after several such separations (Figs. 36 and 38). After separation from the sporophore, but while still in a chain, each aeciospore initial divides into an immature aeciospore and a wedge-shaped intercalary cell (Fig. 38) which resides between each of the aeciospores in the chain. The intercalary cells, sometimes referred to as "disjunctor cells," eventually disintegrate, releasing the mature aeciospores. Although the aeciospore *initials* may be annellophoric, the ultimate cleaving of the aeciospores and intercalary cells is clearly an arthrosporic process as diagnosed by Hughes (1970). A similar account of aeciosporogenesis is provided for *Puccinia graminis* f. sp. *tritici* (Holm and Tibell, 1974). The research by Gold *et al.* (1979) on *Puccinia recondita* emphasized later stages in the developmental

Fig. 27.   The elongated peridial cells (PC) of *Gymnosporangium juniperi-virginianae* are overlapped (imbricate) at their apices (arrows), facilitating the opening and closing of the peridium with changes in humidity. Their surface is covered with finely striated ridges. (× 1,040) (L. J. Littlefield, unpublished.)

Fig. 28.   The apices of aecia (AE) of *Gymnosporangium globosum* shred longitudinally, thereby releasing aeciospores borne in the sori. (× 60) (L. J. Littlefield, unpublished.)

Fig. 29.   The peridermiod aecia (AE) of *Coleosporium senecionis* attain a length of approximately 0.7-1.0 mm on the needles of their host. The peridium remains intact until mature, at which time it ruptures longitudinally, see Fig. 30. (× 40) (L. J. Littlefield, unpublished.)

Fig. 30.   A mature aecium of *Coleosporium senecionis* in which the peridium has ruptured (arrow), allowing the aeciospores to escape. (× 30) (L. J. Littlefield, unpublished.)

Fig. 31.   The peridermioid aecia of *Cronartium ribicola* rupture in numerous, random ways. Here, the peridium (PR) has remained generally intact, having separated along its base. It remains attached at one point (arrow). A, aeciospores. (× 16) (L. J. Littlefield, unpublished.)

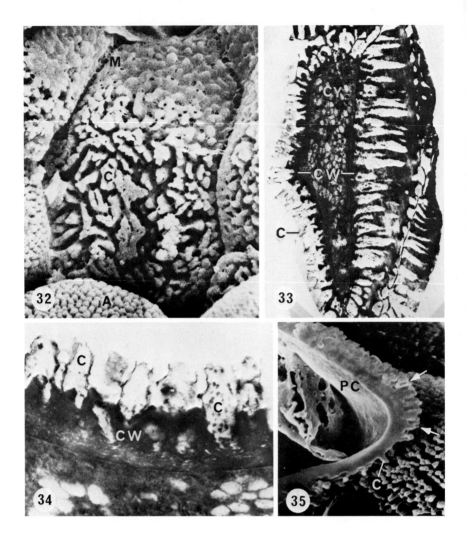

Figs. 32–34. Peridial cells of *Puccinia recondita*. (From Gold *et al.*, 1979. Reproduced by permission of the National Research Council of Canada from *Can. J. Bot.* **57**, 74–86.)

Fig. 32. Surface view showing progressive exposure of surface ornamentation as the interstitial matrix (M) between the clavae (C) disappears. Compare the coarse, irregular ornamentation of the peridial cells with that of the smaller, more regular verrucae on the surface of an adjacent aeciospore (A). (× 4,800)

Fig. 33. Longitudinal section of a peridial cell showing the bizarrely shaped ornaments. The ornaments (dagger-shaped, electron-lucent structures) remain completely embedded in the thick-walled, outer side (solid arrow) of the cell (away from the aeciospore column). On the side toward the aeciospores (open arrow), the cell wall is relatively thin, allowing the apical, truncated portions of the clavae (C) to be exposed. The bases of the clavae are embedded in the cell wall (CW). CY, cytoplasm. (× 5,000)

process, e.g., intercalary cell disintegration, and aeciospore and peridial ornamentation (see Section IV, this chapter). The catenulate nature of immature aeciospores and the wedge-shaped configuration of intercalary cells is clearly evident with scanning electron microscopy (Fig. 39).

In contrast to the thin, wedge-shaped intercalary cells (Figs. 38 and 39) of aecidioid aecia, the intercalary cells of the peridermioid aecia of *Cronartium ribicola* are cylindrical cells 20–30 μm long, more or less equal in length to aeciospores of that fungus, as demonstrated by light microscopy (Colley, 1918). In both cases, however, the intercalary cells are thin-walled, lack ornamentation, and disintegrate prior to aeciospore discharge.

To the authors' knowledge, sporogenesis in uredinoid aecia has been studied only in the case of *Trachyspora intrusa* (Henderson, 1973). That study dealt only with the later stages of spore development, particulary ornamentation. Although catenulate, the uredinoid aeciospores of that rust were shown to have no intercalary cells.

C. Aeciospore Ornamentation*

The surfaces of all aeciospores yet examined by electron microscopy bear some type of ornamentation; none have been found to be entirely smooth. Smooth areas do exist over portions of the aeciospore surface in some species of *Coleosporium* (N. Hiratsuka and Kaneko, 1975) and *Cronartium* (Antonopoulos and Chapman, 1976; Y. Hiratsuka, 1971) (Fig. 42). This property is characteristic of mature spores and is not due to spore immaturity as is a smooth aeciospore surface in *Puccinia recondita* (Fig. 55) (Gold *et al.*, 1979). Ornaments consist of electron-lucent materials distinct from the electron-opaque spore wall. In *Cronartium fusiforme* the ornamental processes are not digested by commercial chitinase as are other components of aeciospore walls (Walkinshaw *et al.*, 1967). The specific types of ornaments can be divided into four categories.

The two most commonly described types of aeciospore ornaments are coglike (Figs. 40, 45, 51, 54) or annulate knobs (Figs. 41, 42, 46, 53). The coglike knobs are more or less cylindrical, have a flattened apex, and are generally 0.3–1.5 μm long. Coglike knobs occur in *Gymnosporangium clavipes* (Kozar and Netolitzky, 1975), *Gymnosporangium cornutum* (von

* Peltate ornaments of a new species *Caeoma peltatum* are described ultrastructurally by C. G. Shaw III (1976). (See Supplementary Bibliography for reference.)

Fig. 34.   Enlargement of a portion of Fig. 33 showing the truncated clavae (C), irregular in outline, and having their bases embedded in the cell wall (CW). (× 17,000)

Fig. 35.   Cross fracture of a peridial cell (PC) of *Puccinia graminis* f. sp. *tritici*. Note the clavae (C), some of which appear to be either branched or partially fused, laterally (arrows). (× 3,800) (L. J. Littlefield, unpublished.)

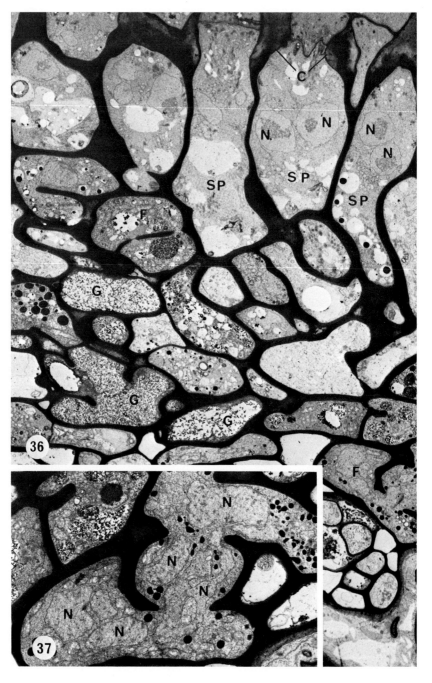

Figs. 36 and 37. Aeciosporogenesis in *Puccinia sorghi.* (Rijkenberg and Truter, 1974b, 1975.)

Hofsten and Holm, 1968; Holm *et al.*, 1970), *Melampsora lini* (Gold and Littlefield, 1979), *Puccinia caricina* (von Hofsten and Holm, 1968; Holm *et al.*, 1970, *P. graminis* f. sp. *tritici* (Holm and Tibell, 1974), *P. poarum* (Henderson *et al.*, 1972b), *P. sorghi* (Rijkenberg and Truter, 1974b), *P. recondita* (Gold *et al.*, 1979), and *Uromyces caladii* (Moore and McAlear, 1961). Similar processes exist in *Gymnosporangium tremelloides* and *Gymnosporangium libocedri*, but they are rather pointed in the former, and shaped somewhat as small, triangular warts in longitudinal sections of the latter (von Hofsten and Holm, 1968; Holm *et al.*, 1970). Closely spaced, conical ornaments (Fig. 43) are typical of *Cronartium comadrae* aeciospores (Y. Hiratsuka, 1971). Large refractile granules occur interspersed among the coglike processes of *Gymnosporangium libocedri* (von Hofsten and Holm, 1968; Holm *et al.*, 1970), *Puccinia caricina*, *P. graminis* f. sp. *tritici* (Figs. 44 and 45), and *P. poarum* (Henderson *et al.*, 1972b). They are absent from *P. recondita* (Fig. 40) (Gold *et al.*, 1979).

Annulate knobs appear in replicas, in longitudinal sections, and in the scanning electron microscope as irregular stacks of usually five to eight discs (Figs. 41, 42, 53), often 1.5–2.5 μm tall. Annulate knobs have been shown in *Coleosporium* spp. (Henderson and Prentice, 1974; Y. Hiratsuka and Kaneko, 1975; von Hofsten and Holm, 1968; Holm *et al.*, 1970), *Cronartium* spp. (Antonopoulos and Chapman, 1976; Grand and Moore, 1972; Y. Hiratsuka, 1970; Mielke and Cochran, 1952; Walkinshaw *et al.*, 1967), and the aecioid teliospores of *Endocronartium harknessii* (Henderson and Hiratsuka, 1974; Y. Hiratsuka, 1970). A somewhat different annulate knob structure occurs in *Chrysomyxa pirolata* aeciospores (Fig. 46). With scanning electron microscopy the knobs appear to consist of only two discs, one on top of the other. A presumably thin covering layer of material appears draped over the ornaments, giving a netlike appearance to the spore surface between the knobs (Y. Hiratsuka, personal communication). Annulate knob ornamentation also characterizes aecioid urediospores of *Coleosporium* spp. (N. Hiratsuka and Kaneko, 1975).

A third type of aeciospore ornamentation, present in *Phragmidium mucronatum*, consists of a combination of large, sharply pointed spines, approximately 1.0 μm long (similar to urediospore spines) interspersed among minute, probably cylindrical projections (Fig. 57) (von Hofsten and Holm 1968; Holm *et al.*, 1970). The large spines are widely spaced, similar to typical urediospore spines rather than the normally closely spaced ornaments

Fig. 36. Section through an aecial stroma showing often-multinucleate fusion cells (F), binucleate sporophores (SP), and cells (G) with degenerate cytoplasm and residue-containing vacuoles. C, collar around distal end of sporophore; N, nuclei. (× 3,000)

Fig. 37. Multinucleate fusion cell. N, nuclei. Some "N" designations may represent different planes of section through the same nucleus. (× 5,000)

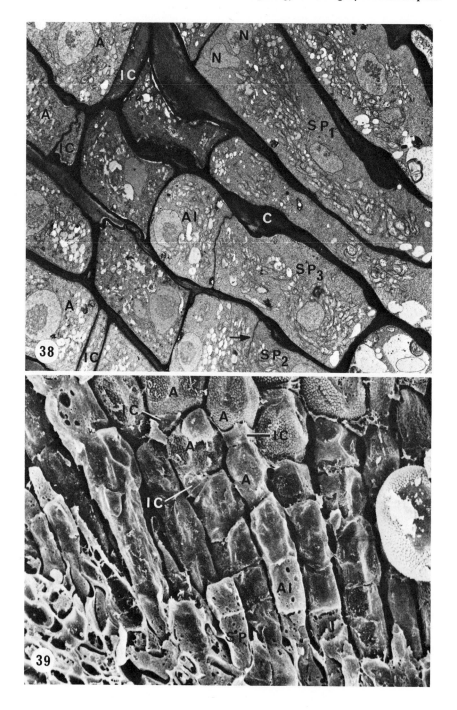

of aeciospores. In *Phragmidium fragariae* the aeciospores are ornamented with smaller, yet pointed spines (0.3 μm tall), which are almost as numerous as peglike knobs in other genera (Henderson and Prentice, 1973). They are borne individually or in clusters of two or three, rarely more, on elliptical to irregular shaped, raised plaques comprised of an electron-lucent material similar to the spines. The minute processes between the spines that occur in *Phragmidium mucronatum* (von Hofsten and Holm, 1968) are absent in *Phragmidium fragariae* (Henderson and Prentice, 1973).

A fourth type of aeciospore ornamentation has been observed on only two *Aecidium* spp. from the South Pacific (Punithalingam and Jones, 1971). The processes of these spores consist of conical, sometimes ribbed, spines which bear conical to hemispherical caps (Fig. 47). These ornaments are somewhat similar to those of certain uredinoid aeciospores described in the following paragraph.

As might be expected, uredinoid aeciospores (Y. Hiratsuka, 1973) share certain ornamentation characteristics with urediospores. Uredinoid aeciospores of *Pileolaria* spp. bear elongated ridges of varying length, similar to many urediospores (Figs. 83–85) in that genus (J. F. Hennen, personal communication). The uredinoid aeciospores of *Trachyspora intrusa* are catenulate, but lack intercalary cells. Their echinulate ornaments are pointed as in urediospores of most genera, but they are often united basally to form raised plaques from which the conical portions arise (Henderson, 1973). The similarities between ornamentation of uredinoid aeciospores and urediospores are especially evident in *Coleosporium* spp. (N. Hiratsuka and Kaneko, 1975). Of particular interest are *Coleosporium evodiae* and *C. phellodendri* in which spines are borne on more or less hemispherical pads, similar to *Trachyspora intrusa* (Henderson, 1973). However, in the two

---

Fig. 38. Section of *Puccinia sorghi* aecium showing stages in the development of aeciospore initials (AI) and aeciospores (A). After conjugate division, daughter nuclei (N) migrate to the apical region of the aeciosporophore (SP$_1$). Centripetal cross-wall formation (arrow, in SP$_2$) results in delineation of aeciospore initial (AI) from the sporophore (SP$_3$). Thickened collars (C) composed of annulations form in the wall of the sporophores as successive aeciospore initials are produced from the sporophores. Aeciospore initials cleave into immature aeciospores (A) and intercalary cells (IC) in more distal portions of the sorus primordium. ($\times$ 3,200) (From Rijkenberg and Truter, 1974b.)

Fig. 39. Fracture through the base of an aecium of *Puccinia recondita;* enlargement of a portion of Fig. 24. The aeciosporophores (SP) give rise to cylindrical aeciospore initials (AI) which divide into alternating aeciospores (A) and intercalary cells (IC). Aeciospore ornamentation becomes progressively evident in distal portions of the spore chains, concomitant with the degradation and crushing (arrow) of the intercalary cells. ($\times$ 1,600) (From Gold and Littlefield, 1979. Reproduced by permission of the National Research Council of Canada from *Can. J. Bot.* **57**, 74–86.)

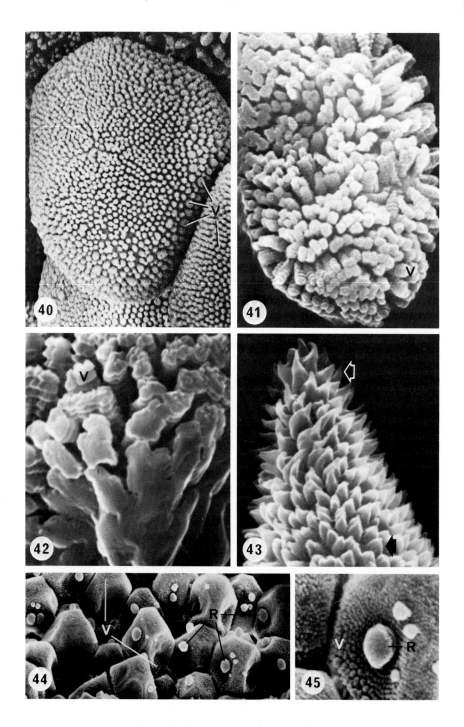

Figs. 40–45. Aeciospore ornamentation.

former species, the spines (structures typical of urediospores) are annulate, a property typical of many aeciospores (N. Hiratsuka and Kaneko, 1975).

A common ontogenetic pattern exists in the formation of aeciospore ornaments and associated refractile bodies that have been studied ultrastructurally (Gold *et al.*, 1979; Holm and Tibell, 1974). Following division of the aeciospore initial into the intercalary cell and the immature aeciospore, the wall of the latter consists of an electron-opaque primary wall originally about 0.1 μm thick. Soon thereafter ornaments begin to form within the primary cell wall (Fig. 48) while the latter is still less than 0.25 μm thick. As the primary wall continues to thicken, the individual ornaments expand in size, apparently due to centripetal addition of their constitutive material. Eventually the thickening of the primary wall and the enlargement of the processes embedded therein ceases, and the fully expanded processes extend from the plasma membrane to slightly below the outer surface of the primary cell wall of the immature aeciospore. Large refractile granules, if present, follow a similar path of development, although their large size often causes a localized invagination of the cell wall and plasma membrane when cells are tightly packed in the aecium (Holm and Tibell, 1974). Upon cessation of primary wall thickening and enlargement of the embedded ornamental processes, the permanent secondary wall is deposited centripetal to the primary wall to a thickness approximately equal to the thickness of the primary wall and the embedded ornaments. Accompanying the deposition of the secon-

Fig. 40. The verrucae (V) of *Puccinia recondita* are small, closely spaced knobs, smooth in outline. (× 3,100) (From Gold *et al.*, 1979. Reproduced by permission of the National Research Council of Canada from *Can. J. Bot.* **57**, 74-86.)

Fig. 41. The verrucae (V) of *Cronartium coleosporioides* are distinctly annulated in outline, appearing similar to a stack of 6-9 discs, with a constriction between each disc. (× 4,000) (From Y. Hiratsuka, 1971. Figures 41-43, reproduced by permission of the National Research Council of Canada from *Can. J. Bot.* **49**, 371-372.)

Fig. 42. Transitional zone between verrucose and smooth areas of *Cronartium coleosporioides* aeciospore. At maturity, both areas exist on the spore; this property is not a characteristic of immaturity in *C. coleosporioides* as it is in *Puccinia* spp. (see Fig. 55). V, verrucae. (× 8,000) (From Y. Hiratsuka, 1971.)

Fig. 43. One end of aeciospores of *Cronartium comandrae* is pointed. Near that end the ornaments are distinctly conical (open arrow) compared to the more bulbous, blunt ornaments (solid arrow) in other regions of the spore. (× 8,200) (From Y. Hiratsuka, 1971.)

Fig. 44. Aeciospores of *Puccinia graminis* f. sp. *tritici* are compressed into angular shapes within the aecium. They are ornamented with minute, knoblike verrucae (V) and larger refractile granules (R). (× 1,400) (L. J. Littlefield, unpublished.)

Fig. 45. Enlargement of a portion of Fig. 44. Note the relative size of the verrucae (V) and the refractile granules (R). (× 4,200) (L. J. Littlefield, unpublished.)

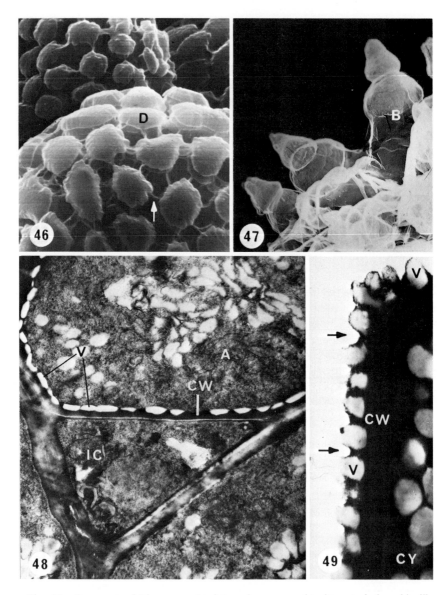

Fig. 46. Ornaments of *Chrysomyxa pirolata* aeciospores consist of two stacked, cushionlike discs (D) with narrow ridges (arrow) interconnecting the stacks. (× 2,700) (Courtesy of Y. Hiratsuka, unpublished.)

Fig. 47. Aeciospores of *Aecidium balansae* are ornamented with conical projections supported on raised, bulbous bases (B). (× 13,000) (From Punithalingam and Jones, 1971. Reproduced by permission of Cambridge University Press from *Trans. Br. Mycol. Soc.* **57,** 325-331.)

dary wall layer is the digestion, reabsorption, or otherwise disappearance of the electron-opaque primary wall layer from around the embedded ornaments (Figs. 49–51). This digestion continues until all the primary wall disappears, or in some cases a thin superficial layer persists, leaving the ornamental processes standing free on the surface of the cell, with essentially all intervening primary wall material absent (Fig. 51). In *Melampsora lini* what apparently is the disintegrating primary cell-wall layer appears as a gelatinous matrix which condenses, or otherwise disappears, to expose the ornaments (Gold and Littlefield, 1979) (Fig. 54). Similar processes are apparent in *Puccinia recondita* (Fig. 55) (Gold *et al.*, 1979).

The knobs on mature *Puccinia recondita* aeciospores reside in small depressions on the spore surface. Exactly when these depressions form during sporogenesis is not known. The knobs are rather loosely attached to the spore surface as they are commonly dislodged during cryofracturing, exposing the underlying depressions in the spore surface (Gold *et al.*, 1979) (Fig. 52).

A diagrammatic summary of the formation of aeciospore ornamentation based on the studies of Holm and Tibell (1974), Gold *et al.* (1979), and Moore and McAlear (1961) is given in Fig. 58. A somewhat modified system of development exists in *Phragmidium fragariae* where the spine-shaped ornaments are subtended by electron-lucent plaques, although the processes of sequential deposition and reabsorption are similar (Henderson and Prentice, 1973). It has been suggested by Gold *et al.* (1979) that digestion of the primary cell wall of aeciospores may result in part from the release of enzymes upon the crushing of the intercalary cells within the aeciospore chain. Primary cell-wall digestion occurs in *Puccinia recondita* only in aeciospores that are located at the same level or distal to crushed intercalary cells.

D. Aeciospore Germ Pores

The germ pores of most aeciospores, examined to date ultrastructurally, have localized interspersions of electron-lucent materials within the otherwise

---

Figs. 48 and 49. Early stages in development of aeciospore ornamentation in *Puccinia recondita.* (From Gold *et al*, 1979. Reproduced by permission of the National Research Council of Canada from *Can. J. Bot.* **57**, 74–86.)

Fig. 48. Longitudinal section through an aeciospore (A) and an intercalary cell (IC). Aeciospore verrucae (V) begin as small, electron-lucent regions within the primary cell wall (CW) adjacent to the plasma membrane. (× 12,500)

Fig. 49. A later stage in which additional cell wall (CW) has been laid beneath the verrucae (V). The original cell-wall matrix between the ornaments (see Fig. 48) has begun to erode, resulting in gaps (arrows) which expose the distal portions of the verrucae. CY, spore cytoplasm. (× 20,800)

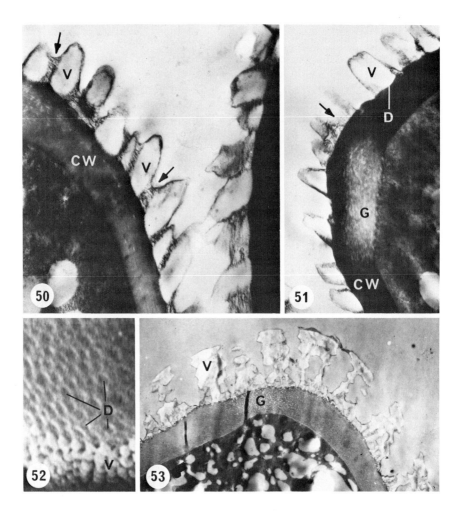

Figs. 50–52. Later stages in development of ornamentation of *Puccinia recondita* aeciospores. (Gold *et al.*, 1979. Reproduced by permission of the National Research Council of Canada from *Can. J. Bot.* **57**, 74–86.)

Fig. 50. The mass of the cell wall (CW) is composed of a secondary layer. The interstitial matrix of the primary cell-wall layer remains only as eroded lamellae (arrows) between the verrucae (V) which are almost entirely exposed. (× 27,500)

Fig. 51. At maturity the verrucae (V) reside on the outer surface of the secondary cell wall, (CW), in small depressions (D), with only small residual fragments (arrow) of the original, interstitial primary wall matrix left between the ornaments. Note the thickened, but electron-lucent germ pore (G) region in the cell wall. (× 22,900)

Fig. 52. Surface view showing verrucae (V) on spore surface and the small depressions in which they reside. These depressions were made visible by dislodging the ornaments during the process of fracturing. (× 10,100)

Fig. 53. Cross section of aeciospore of *Coleosporium tussilaginis* showing the stacks of discs comprising the verrucae (V) on the spore surface, G, germ pore. (× 6,000) (From von Hofsten and Holm, 1968.)

electron-opaque spore wall. In some cases, e.g., *Gymnosporangium cornutum* and *Gymnosporangium tremelloides,* the spore wall is not altered in thickness within the germ pore, but the pores are surrounded by an annular thickening compared to the remainder of the spore wall (Fig. 56) (von Hofsten and Holm, 1968). An annular thickening is absent from the periphery of aeciospore germ pores in *Gymnosporangium libocedri* (von Hofsten and Holm, 1968) and *Trachyspora intrusa* (Henderson, 1973). The germ pores of *Phragmidium mucronatum* (van Hofsten and Holm, 1968) and *P. fragariae* (Henderson and Prentice, 1973) contain a greatly thickened, lens-shaped, alveolate mass of wall material (Fig. 57) which is somewhat more electron-opaque than the rest of the cell wall. The germ pore region of *Puccinia sorghi* (Rijkenberg and Truter, 1974b) and *Puccinia recondita* (Gold *et al.,* 1979) (Fig. 51) aeciospores is only slightly thickened. In *Cronartium fusiforme* (Walkinshaw *et al.,* 1967) and *Coleosporium tussilaginis* (Henderson and Prentice, 1974; von Hofsten and Holm, 1968) the germ-pore region consists of a locally thin area containing electron-lucent interspersions compared to the rest of the spore wall (Fig. 53).

Mild treatment with chitinase digests the wall substance within the germ pore of *Cronartium fusiforme* aeciospores and allows the protoplast to flow out. Upon normal germination, the germ-tube wall is not continuous with any discernible layer of the aeciospore wall and appears to be synthesized concurrently with passage of the cytoplasm from the spore into the germ tube (Walkinshaw *et al.,* 1967).

Typically the germ tubes from aeciospores form appressoria and penetrate their hosts by growing through stomata. An exception to this is direct penetration of host epidermal cells by infection pegs formed from appressoria on host surfaces by *Gymnoconia interstitialis* (Pady, 1935a).

E. Ontogeny of Peridium Ornamentation

As discussed in Section IV,A,3, this chapter, the most extensive research on peridium ornamentation is based on light microscopy of rostelioid aecia. However, recent studies by Gold *et al.* (1979) showed that ornaments on the peridial surface of *Puccinia recondita* form in a similar, but not the same, manner, as do those on aeciospores. Although the electron-lucent, clavate projections become progressively exposed by the dissolution of interstitial cell wall material, at maturity approximately half the length of each clava remains embedded within the spore wall (Fig. 34). Clavae form on all sides of the spore, but are exposed only on the side which undergoes partial dissolution of the cell wall. On the thick-wall side of the peridial cell the dagger-shaped projections remain totally embedded in the cell wall (Fig. 33). When fully exposed, the peridial ornaments are more widely and irregularly spaced and taller than aeciospore ornaments (Fig. 32).

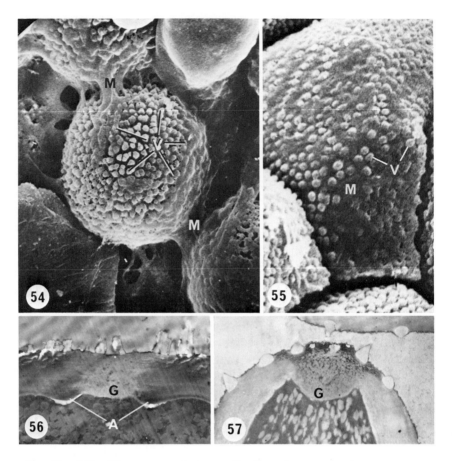

Figs. 54 and 55. Disappearance of primary cell-wall matrix to reveal aeciospore ornaments.

Fig. 54. The interstitial matrix (M) between verrucae (V) of *Melampsora lini* aeciospores is mucilagelike and appears to condense or otherwise digest, exposing the ornaments. (× 2,400) [From Gold and Littlefield, 1979. Reproduced by permission of the National Research Council of Canada from *Can. J. Bot.* (in press).]

Fig. 55. A similar process occurs in *Puccinia recondita*, although the interstitial matrix (M) lacks the mucilagelike appearance of that in *M. lini*. V, verrucae. × 6,700. (From Gold *et al.*, 1979. Reproduced by permission of the National Research Council of Canada from *Can. J. Bot.* 57, 74–86.)

Figs. 56 and 57. Aeciospore germ pores. (From von Hofsten and Holm, 1968.)

Fig. 56. The aeciospore germ pore (G) in *Gymnosporangium cornutum* is a comparatively electron-lucent region having about the same thickness as the rest of the spore cell wall, but is surrounded by a slightly thicker annulus (A) of cell wall material. (× 5,400)

Fig. 57. The germ pore (G) in *Phragmidium mucronatum* is a pluglike structure, considerably thicker than the surrounding cell wall, and contains interspersions of electron-opaque materials. (× 7,500)

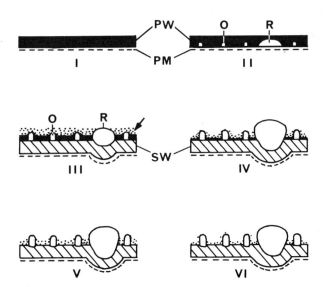

Fig. 58. Diagrammatic summary of the development of aeciospore ornaments and refractile granules, the latter structures being restricted to only certain species. Modified from Holm and Tibell, 1974. (I) PM, plasma membrane; PW, original primary wall; so labeled in all stages. (II) Ornaments (O) (typically are verrucae) begin to develop as deposits of electron-lucent materials within the primary wall (PW). The large electron-lucent, i.e., white, area within the primary wall represents the initiation of a refractile granule (R). (III) A secondary layer (SW) of wall material is laid down beneath the primary layer (PW) in which ornaments (O) occur. The ornaments continue to increase in size, and the outer portions (stippled area; arrow) of the primary wall begin to erode. R, refractile granuler. (IV) The ornaments, including the refractile granule, have reached their maximum size, and the secondary wall layer has reached its maximum thickness, approximately. The primary wall layer continues to degrade, becoming thinner and exposing the ornaments to a greater depth. (V) and (VI) Progressive disappearance of the primary wall matrix between the ornaments and the refractile granule, leaving only a small residue (stippled area) on the surface between the ornaments. The secondary wall surface is slightly depressed under the individual ornaments and is invaginated into the cell lumen beneath the large refractile granules.

F. Aeciospore Dehiscence

In addition to passive dislodging of aeciospores resulting from wind and physical movement of host tissue, a forcible discharge mechanism exists that can propel aeciospores of numerous rusts to a height of 4–15 mm (Buller, 1958). Buller suggested that this mechanism results from two opposing forces (a) the adhesion of spores to one another due to the cementing intercellular matrix (see Fig. 54), and (b) turgor pressure of the cells which tends to inflate the spores spherically from their initial, compressed, angular shape (see Fig. 44). As the turgidity increases, and at the moment when the adhesive force is overcome by the expansive force, the result is the sudden and violent discharge of the terminal spore of the chain. The intermeshing ornaments of adjacent spores may be involved in maintaining the adhesive force until the critical moment when increasing turgor pressure suddenly overcomes that force. The role, if any, of the refractile granules is difficult to assess since they are present, e.g., *Puccinia graminis* f. sp. *tritici* (Figs. 44 and 45), or absent, e.g., *P. recondita* (Fig. 39), in closely related rusts which both exhibit forcible discharge of aeciospores. Studies with *Gymnosporangium myricatum* suggest that refractile granules are responsible, in part, for the sudden, forcible discharge of individual aeciospores, likened to "popped corn jumping out of the popper" (Dodge, 1924). Combined experimental and ultrastructural research would be a most desirable approach to studying this question of "aecidial gunnery," to borrow a term from Buller (1958).

An interesting relationship exists between the pattern of cell-wall thickening of peridial cells in aecidioid and roestelioid aecia and the respective response of the peridia in these aecial types to changes in humidity. As already described (see Section IV,A,2, this chapter), the outer walls of peridial cells of aecidioid aecia are thickened ("outer" meaning the side of the peridial cells away from the aeciospores, "inner" meaning the side toward the aeciospores). Apparently, as in the annulus of fern sporangia (Foster and Gifford, 1974), the thickened portion of the cell wall is less extensible and thus responds less to changes in moisture than the thin portion of the wall of the same cell. Consequently, when exposed to low humidity the thin, inner walls of peridia of aecidioid aecia contract and keep the peridial covering intact over the chains of aeciospores (Fig. 23). Upon absorption of water the thin, inner walls of the peridial cells expand more than do the thick, outer cell walls, causing the rupture of the peridial layer (Fig. 22) and the subsequent release of aeciospores. Due to the pressure built up by the expanding chains of aeciospores and (possibly) the compression of dense refractile granules between adjacent spores (Fig. 44), the sudden rupture of the peridium causes the aeciospores to be forcibly ejected over a distance of several millimeters (Buller, 1950, 1958).

When roestelioid aecia are kept moist, the thin, outer walls expand more than do the thick, inner walls causing the files of shredded peridial cells to straighten and form an enclosure over the chains of aeciospores (Pady *et al.*, 1968). Upon drying, the thin, outer walls shrink, causing the recurved opening of the peridial segments, thus exposing the aeciospores to the air (Fig. 25). The positive correlation between reduction in ambient humidity and enhanced aeciospore release has been demonstrated in *Gymnosporangium juniperi-virginianae* (Pady *et al.*, 1968). A similar relationship between aeciospore dehiscence under dry conditions and not under wet conditions, as controlled by hygroscopic swelling of the thin, outer cell walls of peridial cells in the balanoid aecia of *Gymnosporangium fuscum* was noted by Leppik (1977). Apparently the surface ornamentation of peridial cells plays little role, if any, in hygroscopic movements of the peridium as that tissue in both aecidioid and roestelioid aecia possesses surface ornamentation on the side of cells facing the inside of the aecium, yet those aecia respond in an opposite manner to moisture. The ornaments might, however, facilitate spore dispersal by preventing the adherence of spores to the peridial surface.

V. UREDIA AND UREDIOSPORES

Uredia are sori which produce vegetative, repeating, dikaryotic urediospores usually on a dikaryotic mycelium (Y. Hiratsuka, 1973). Uredia form on leaves, stems, fruits, or fronds of herbaceous plants. No rusts, to the authors' knowledge, produce uredia on woody stems or branches as commonly happens with pycnia and aecia in perennial infections, e.g., *Cronartium* spp. Typically uredia are formed subepidermally, becoming erumpent (Fig. 59). Several exceptions occur, however. For example, *Hemileia* spp. and other superstomatal uredial forms produce pedicellate urediospores on clavate, sporogenous cells that extend beyond the leaf surface through stomata (Fig. 60). The mature uredia of certain rusts are covered by a peridium, a one-cell-thick layer which resides immediately below the epidermis of the host. It is a somewhat hemispherical structure in such rusts, extending over the entire sorus. The peridium may open by irregular fissures, e.g., *Milesia* spp. or it may have a distinct ostiole, e.g., *Pucciniastrum* spp. (Fig. 61) (Arthur, 1929; Cummins, 1959). A rudimentary, simple peridium consisting of a single layer of smooth, thick-walled, more or less isodiametric cells covers the immature uredia of *Melampsora lini* (Fig. 62), and is shown best by light microscopy. This layer of cells, closely appressed to the host epidermis, is ruptured and sloughed off during expansion of the uredium and persists as fragments, united with ruptured epidermis, around the margin of mature uredia (Fig. 59) (Hassan and Littlefield, 1979). Peridia are typically absent from uredia of *Puccinia* spp. and *Uromyces* spp. (Cummins, 1959).

From light microscopy of 112 species in 65 genera, Kenney (1970) distinguished 14 types of uredia, based on (a) the absence or presence, and the nature of accessory structures (paraphyses and peridia), (b) the position of the hymenium in host tissue, and (c) the lateral extent of the hymenium. A similar ultrastructural comparison of these uredial types is yet to be made.

Urediospores (Figs. 60, 69–71) are typically one-celled and pedicellate, except for catenulate, aecioid urediospores, e.g., *Chrysomyxa* spp. and *Coleosporium* spp. Most commonly, urediospores are round to ovate (Figs. 61, 80, 84, 85), although in *Uredinopsis* spp. they are lanceolate to fusiform with a spine-like beak (Fig. 81), reniform in *Hemileia* spp. (Figs. 60 and 78), and irregularly obovoid to clavate in *Milesia* spp. (Fig. 82). Although round in

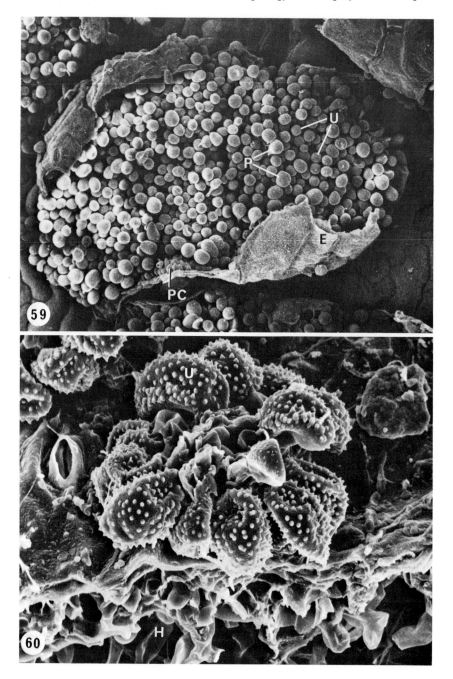

outline, the urediospores of *Uromyces phaseoli* var. *typica**** (Hardwick *et al.*, 1975) are shaped similar to a doughnut. Although such a configuration is typical of dehydration damage, i.e., an artifact, Hardwick *et al.* (1975) contend that such a shape does occur in nature as evinced by light microscopy of dry spores. The sunken areas on opposite sides of the spore, visible by light microscopy, are the sites of germ pores where the walls are thicker and surface ornamentation is reduced. Several possible survival advantages of this shape compared to a sphere were discussed by Hardwick *et al.* (1975). Thick-walled, darkly pigmented urediospores capable of enduring long periods of unfavorable conditions occur in some species of *Puccinia* and *Hyalopsora*. Such spores, known as amphispores, have not been studied ultrastructurally. Urediospores are absent from the life cycle of certain rusts, e.g., many *Gymnosporangium* spp.

In addition to urediospores, the uredium may contain sterile, smooth, clavate to capitate or basally united paraphyses (Cummins, 1959). Paraphyses may be restricted to the periphery of the sorus, e.g., *Phragmidium* spp., or scattered among the urediospores, e.g., *Melampsora* spp. (Figs. 68 and 71).

A. Urediospore Formation

Uredial ontogeny has been documented ultrastructurally in *Puccinia coronata* f. sp. *avenae* and *Puccinia graminis* f. sp. *avenae* (Harder, 1976c), *Uromyces phaseoli* var. *typica* (Müller *et al.*, 1974c), *Puccinia sorghi* (Rijkenberg, 1975), and *Melampsora lini* (Hassan and Littlefield, 1979). The terminology used by Harder (1976c) is accepted by the present authors, based in part on a precedent set by Hughes (1970), who, by light microscopic study showed urediospores to be sympoduloconidia, a conclusion confirmed by Harder (1976c).

---

* The bean rust fungus is variously known as *Uromyces appendiculatus*, *U. phaseoli*, and *U. phaseoli* var. *typica*. To avoid confusion, and distinguish it from the closely related, but different, *U. phaseoli* var. *vignae* (cowpea rust), we will use *U. phaseoli* var. *typica* throughout this book regardless of the name used by the author(s) cited.

Fig. 59. Surface view of a uredium of *Melampsora lini*. The uredium contains relatively large, sterile, capitate paraphyses (P) and smaller urediospores (U). The sorus is surrounded by fragments of ruptured host epidermis (E), to the underside of which are adhered the peridial cells (PC) of the fungus. (× 250) [From Hassan and Littlefield, 1979. Reproduced by permission of the National Research Council of Canada from *Can. J. Bot.* (in press).]

Fig. 60. Surface view of a superstomatal uredium of *Hemileia vastatrix*. Note the reniform urediospores (U) and the interior of the host leaf (H) exposed by cutting the tissue. (× 1,100) (L. J. Littlefield, unpublished.)

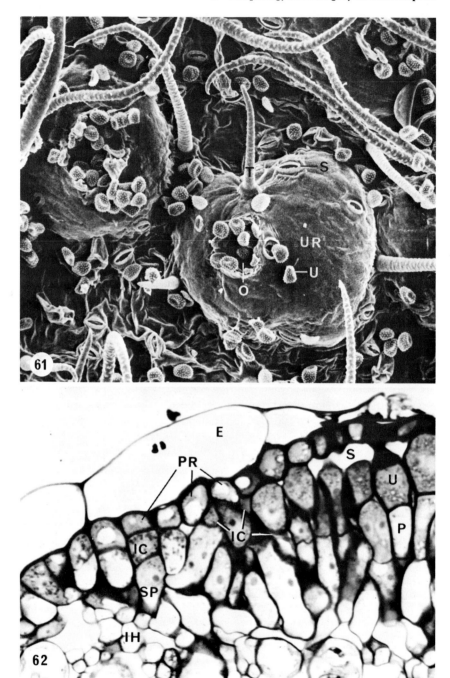

The fundamental cell in urediosporogenesis is the sporogenous cell (Figs. 63 and 67), many of which occur in each uredium. This is a hypha-like cell, usually swollen at one end, and tightly bound to adjacent cells by an electron-opaque intercellular matrix. The sporogenous cell gives rise, by budding, to sympodially produced, elongated spore buds (Fig. 63). Conjugate nuclear division of the dikaryotic sporogenous cell provides both the sporogenous cell and the spore bud with a pair of nuclei. Following the conjugate nuclear division a septum forms near the base of the elongated spore bud, separating the latter (having then become the urediospore initial) from the sporogenous cell (Fig. 64). This separation is followed by another conjugate nuclear division and the formation of a second septum which separates the urediospore initial into a proximal pedicel and a distal, immature urediospore (Fig. 65) (Harder, 1976c; Hassan and Littlefield, 1979). Urediospore maturation is accompanied by the development of surface ornamentation (see Section IV,B, this chapter).

The above description of urediospore development from uredial sporogenous cells is essentially the same for Pucciniaceae and Melampsoraceae that have been examined. In some Melampsoraceae additional complexity is added by the formation of certain accessory cells, e.g., peridial cells, intercalary cells, and paraphyses which are typically absent in uredia of Pucciniaceae. The ultrastructural manifestations of the development and demise of these sometimes ephemeral structures may occur before, after, or simultaneously with the development of urediospores from sporogenous cells as described above (Hassan and Littlefield, 1979).

The first evidence of uredium formation in *Melampsora lini* is the aggregation of hyphae to form a uredial initial, usually in a substomatal cavity. The hyphae of the uredial initial orient vertically into a palisade beneath the epidermis and divide to form a basal cell and a terminal cell (Fig. 66). Each terminal cell divides again to form a distal peridial cell and a proximal intercalary cell (Fig. 67); the basal cell of each of the three-celled-columns

---

Fig. 61.   Surface view of uredia (UR) of *Pucciniastrum agrimoniae*. The uredium is enclosed by a persistent peridium beneath the host epidermis, both structures being extended over the surface of the uredium. Urediospores (U) are released through an ostiole (O) in the peridium/epidermis complex. S, host stoma; T, host trichome. (× 380) (L. J. Littlefield, unpublished.)

Fig. 62.   Cross section, light micrograph of immature uredium of *Melampsora lini*. The host epidermis (E) is still intact, and the fungal peridium (Pr) exists as a single layer of cells beneath the epidermis. Urediospores (U), yet immature, are attached to pedicels (P) near the center of the sorus, that area also containing spaces (S) formed by the degradation of ephemeral intercalary cells (IC). Uredia expand radially, thus progressively younger stages of development are evident toward the periphery of the sorus. IH, intercellular hyphae; SP, sporogenous cell. (× 1,360) [From Hassan and Littlefield, 1979. Reproduced by permission of the National Research Council of Canada from *Can. J. Bot.* (in press).]

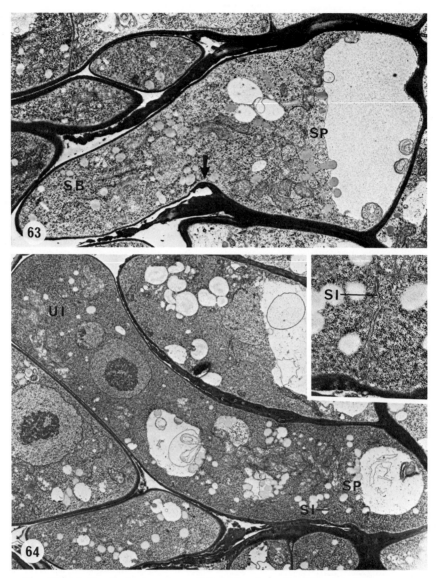

Figs. 63 and 64. Early stages in urediosporogenesis in *Puccinia coronata* f. sp. *avenae*. (From Harder, 1976c. Reproduced by permission of the National Research Council of Canada from *Can. J. Bot.* **54,** 1010-1019.)

Fig. 63. A sporogenous cell (SP) has given rise to a spore bud (SB) which appears to have formed by outgrowth of the inner wall through a rupture in the outer wall layer of the sporogenous cell (arrow). (×6,000)

Fig. 64. A later stage, where the spore bud has elongated to form a urediospore initial (UI). A septum initial (SI; see inset also) has formed and will eventually cleave the urediospore initial from the sporogenous cell (SP). (×5,730; Inset ×23,000)

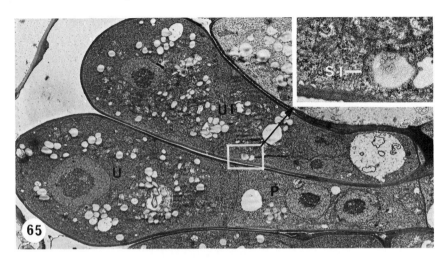

Fig. 65.   Later stages of urediosporogenesis in *Puccinia coronata* f. sp. *avenae* compared to Fig. 64. The urediospore initial (UI) has formed a septum initial (SI; see inset also) which will eventually cleave the urediospore initial into an immature urediospore (U) and a pedicel cell (P). Such a separation has already occurred in the lower pair of the two structures illustrated here. (×4,400; Inset ×12,800) (From Harder, 1976c. Reproduced by permission of the National Research Council of Canada from *Can. J. Bot.* **54**, 1010-1019.)

becomes a sporogenous cell. The middle, or intercalary, cell of each column is short lived, soon becoming crushed and disintegrated during expansion of the uredium. The sporogenous cells bud sympodially, in a manner similar to that described above for *Puccinia coronata* f. sp. *avenae,* and likewise proceed to form urediospore initials which are cleaved into pedicels and immature urediospores. Upward expansion of the uredium and accompanying rupture of the epidermis and peridium are provided for by the enlarged paraphyses which extend above the developing urediospores and push up the overlying structures (Fig. 68). Disappearance of the ephemeral intercalary cells provides room for expansion of the urediospores within the yet unruptured, immature uredium (Figs. 62 and 68). The peridial cells remain attached to the overlying epidermis and persist as a layer of cells on the inner face of the host epidermis that surrounds the ruptured, mature uredium (Figs. 59 and 71). As the urediospores mature, the pedicels eventually separate from the urediospores (Figs. 69 and 70).

The formation of peridial and intercalary cells in *Melampsora lini* which accompany the formation of the first "generation" of urediospores is not repeated in subsequent generations of urediospores produced in the sorus. The formation of the latter follows the basic pattern described for *Puccinia* spp. Uredium ontogeny progresses in a radial manner, with progressive stages of maturity extant from the center to the periphery of an individual sorus

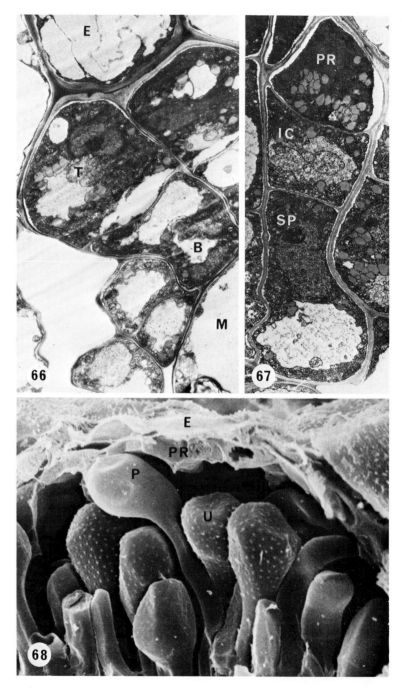

Figs. 66 and 67. Early stages in the formation of a uredium of *Melampsora lini*. [From Hassan and Littlefield, 1979. Reproduced by permission of the National Research Council of Canada from *Can. J. Bot.* (in press).]

(Fig. 62) (Hassan and Littlefield, 1979). The paraphyses are not ephemeral as are peridial and intercalary cells, but remain in the uredium during its entire development. Presumably they do not continue to form past the first generation of urediospores as they are not released from the uredium and their numbers do not increase during the functional life of the uredium.

Although not studied ultrastructurally, a most unusual mode of urediosporogenesis occurs in the tropical rusts, *Kernkampella breyniae-patentis* (Rajendren, 1970) and *Intrapes palisadum* (Hennen and Figueiredo, 1979). Endogenously produced urediospores are formed by proliferation of the sporogenous cells through the apically flared remnants of the pedicels of previous urediospores. Those remnants form a series of collars which subtend the urediospores. Such rusts would provide exceptionally interesting objects for ultrastructural study.

Peculiar inclusion bodies occur in paraphyses, pedicel cells, intercalary cells, and other nondesignated cells of the uredial sorus of *Puccinia coronata* f. sp. *avenae* (Harder, 1976c) and *Melampsora lini* (Hassan and Littlefield, 1979). They are variable in size and shape, including apparently spherical, discoid, reniform, sickle-shaped, rodlike, and other configurations (Fig. 72). Serial sections of these inclusions have not been made; thus the various shapes might represent simply different planes of section through structures not greatly different in overall configuration. The inclusions have a most striking lamellar, internal structure, with a repeating interval of approximately 6–8 nm between lamellae (Figs. 73 and 74). Although no cytochemical investigations have been made, the inclusions closely resemble certain lipidic inclusions in teliospores of *Tilletia caries* (Gardner and Hess, 1977). Similar inclusions also occur in pycnial cells (see Fig. 4 of Mims *et al.*, 1976) and telial buffer cells (see Fig. 2 of Mims, 1977b) of *Gymnosporangium juniperi-virginianae*. The striated appearance of these inclusions is similar to the interstitial material between spores in immature areas of the uredial sorus of *Melampsora lini* (Figs. 75 and 76).

A second type of inclusion in *Puccinia coronata* f. sp. *avenae* paraphyses is a fine membranous network arranged in a "crochet-like" pattern (see Fig. 18

Fig. 66. The uredial initial cells which form beneath the host epidermis (E) first divide into two, forming a terminal (T) and a basal (B) cell. M, host mesophyll cell. ($\times 4,500$)

Fig. 67. The terminal cell (see Fig. 66) of the uredial initials divides again to form a peridial (PR) cell and an intercalary (IC) cell. The basal cell (see Fig. 66) becomes the sporogenous (SP) cell which will subsequently give rise to urediospores. ($\times 4,750$)

Fig. 68. Late in uredium ontogeny of *Melampsora lini*, but before rupture of the sorus, the paraphyses (P) extend above the urediospores (U), pushing up the overlying peridium (PR) and crushed host epidermis (E). ($\times 1,800$) [From Hassan and Littlefield, 1979. Reproduced by permission of the National Research Council of Canada from *Can. J. Bot.* (in press).]

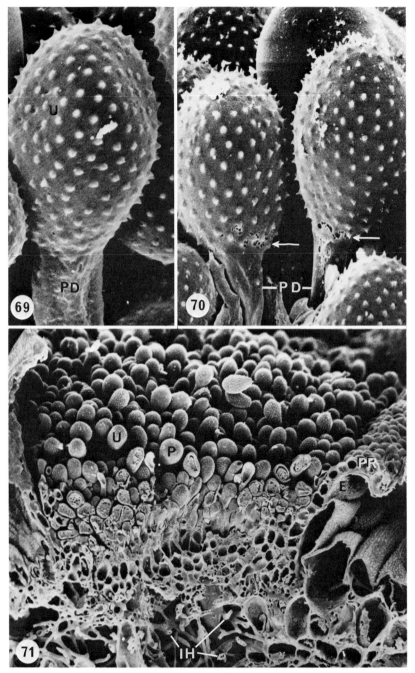

Figs. 69 and 70. Maturation and release of urediospores of *Melampsora lini.* [From Hassan and Littlefield, 1979. Reproduced by permission of the National Research Council of Canada from *Can. J. Bot.* (in press).]

of Harder, 1976c); this structure also occurs in unidentified portions of the uredium of *Uromyces phaseoli* var. *typica* (M. C. Heath, unpublished).

The cross wall separating the urediospore from its pedicel is differentially thickened. The proximal side (the apex of the sporophore) is approximately the same thickness as ordinary cross walls (see Chapter 4, Section I,A), but the distal side (the base of the urediospore) is thickened approximately equal to the rest of the urediospore wall. A small-diameter channel extends from the lumen of the spore through the thickened wall to the septal pore of *Puccinia graminis* f. sp. *tritici*. The diameter of the channel is approximately equal to that of the central region of the septum around the pore which tapers abruptly in thickness (see Fig. 9 of Ehrlich and Ehrlich, 1969). The pore is plugged with an electron-opaque substance similar to that of other septal pores (Littlefield and Bracker, 1971b; Jones, 1973).

B. Urediospore Ornamentation

With few exceptions urediospores are echinulate, bearing on their surface conical spines sometimes curved near their tips (Figs. 70, 86, 87, 101). However, a mutant of *Melampsora lini* exists which is finely verrucose and has small, rounded warts in place of spines (Fig. 77) (Littlefield, 1971b). Commonly, echinulate ornaments are distributed over the entire urediospore surface, although in *Hemileia* spp. the typically asymmetrical urediospores have a smooth, flat, or concave side and an echinulate, convex side (Fig. 78). The conical spines on the surface of *Coleosporium phellodendri* are subtended by a more or less hemispherical, cushion-shaped mass of wall material (N. Hiratsuka and Kaneko, 1975). Those masses may unite laterally over portions of the spore to form a continuous, although undulating, surface below the spines. Such variants may be reflections of different stages of spore maturity, however, as shown in *Puccinia sparganioides* (Amerson and Van Dyke, 1978). Aecioid urediospores of several species of *Coleosporium* and

---

Fig. 69. The base of the urediospore (U) is attached to the apex of the pedicel (PD). The pedicel has a wide diameter and shows little shriveling, both evidence of a urediospore that is not fully mature. ($\times 3,700$)

Fig. 70. In a later stage of development than that shown in Fig. 69, the pedicels (PD) become shriveled in appearance and begin to separate (arrows) from the urediospores ($\times 3,000$)

Fig. 71. Cross fracture of a mature uredium of *Melampsora lini*. The echinulate urediospores (U) and the smooth paraphyses (P) are borne from basal sporogenous tissue (SP) in the sorus. The layer of peridial cells (PR) remains attached to the crushed, ruptured, host epidermis (E) around the periphery of the sorus. IH, intercellular hyphae. ($\times 440$) (From Hassan and Littlefield, 1979. Reproduced by permission of the Research Council of Canada from *Can. J. Bot.* (in press).]

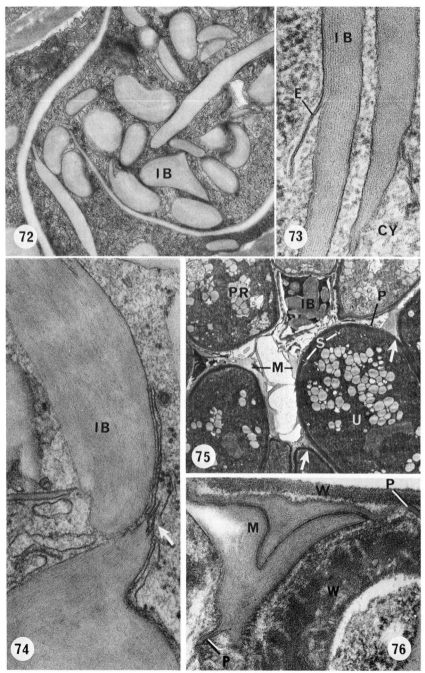

Figs. 72–74. Characteristic inclusion bodies in intercalary cells, some paraphyses, and other cells of *Melampsora lini* uredia. [From Hassan and Littlefield, 1979. Reproduced by permission of the National Research Council of Canada from *Can. J. Bot.* (in press).]

Chrysomyxa are typically verrucose (N. Hiratsuka and Kaneko, 1975) (Fig. 79), similar to aeciospores of those genera. Verrucae, as well as spines, may not be uniformly distributed over the urediospore surface. For example, verrucae of *Puccinia colata* are restricted to the ends, with the central portions of the spore surface being smooth (Fig. 80). Urediospores of *Milesia laeviuscula* are smooth (Fig. 82) and those of some *Uredinopsis* spp. (Fig. 81) are smooth except for two finely verrucose to serrate, longitudinal ridges (Arthur, 1934; Ziller, 1974). Numerous longitudinal ridges of varying length, curvature, and composition occur on the surface of *Pileolaria* spp. urediospores (Figs. 83–85).

Urediospores of *Puccinia graminis* f. sp. *tritici* have 200–300 spines (Cassell *et al.*, 1950) and *Melampsora lini* have at least 200. Spines in the latter are sometimes arranged in a somewhat helical pattern around the longitudinal axis of the spore (Littlefield, 1971b). Standbridge and Gay (1969) were unable to separate four races of *Puccinia striiformis* on the basis of surface features. Silhouette images of whole spores (Pavgi and Flangas, 1965; Rapilly, 1968) lacked sufficient resolution to distinguish more than the general outline of spines.

In thin section urediospore spines are electron-lucent compared to the electron-opaque spore wall (Fig. 100). The spines contain longitudinally oriented, fibrillar materials (Fig. 92) (Amerson and Van Dyke, 1978; Harder, 1976c; Hardwick *et al.*, 1975; Littlefield and Bracker, 1971b;

Fig. 72. Low magnification showing the diversity of shapes which the inclusion bodies (IB) exhibit, and also their large number per cell. ($\times$ 18,900)

Fig. 73. The inclusion bodies (IB) consist of striated materials having 6–8 nm spacing between the strata. Commonly a flaplike extension (E) of two of the outermost membranelike layers of the inclusion extend into the ground cytoplasm (CY) of the cell. ($\times$ 120,000)

Fig. 74. Often near the periphery of the inclusion bodies (IB) numerous membranelike configurations are present (arrow). ($\times$ 84,000)

Figs. 75 and 76. Cross sections within an immature uredium showing the similarity among the inclusion bodies, the intercellular matrix, and the pellicle surrounding urediospores.

Fig. 75. An intercalary cell containing inclusion bodies (IB) is partially crushed. A matrix (M) occurs in the intercellular spaces and is in close association (arrows) with the pellicle (P) which invests the urediospores (U). PR, peridial cells; S, partially developed spines of urediospores. ($\times$ 5,000) [From Hassan and Littlefield, 1979. Reproduced by permission of the National Research Council of Canada from *Can. J. Bot.* (in press).]

Fig. 76. Detail of one intercellular interstice showing the lamellar nature of the intercellular matrix (M) and its continuity with the urediospore pellicle (P). W, cell wall. ($\times$ 72,000) (From Littlefield and Bracker, 1971b. Reproduced by permission of the National Research Council of Canada from *Can. J. Bot.* **49**, 2067–2073.)

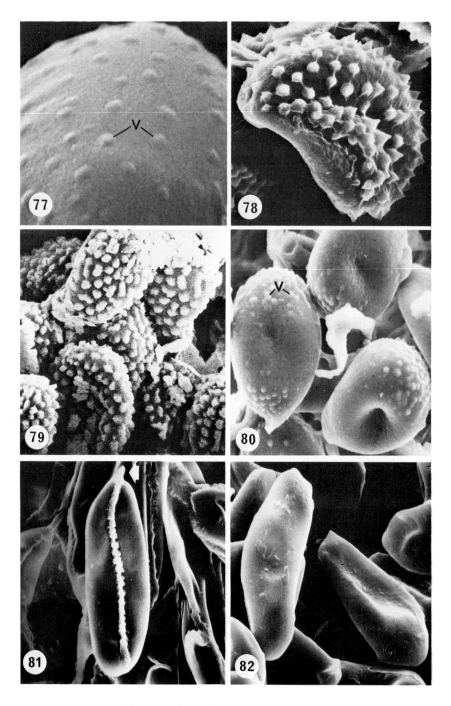

Figs. 77–82. Variations in urediospore ornamentation.

Müller *et al.*, 1974a), and in some cases have an electron-opaque core (Hardwick *et al.*, 1975; Müller *et al.*, 1974a).

In several *Puccinia* spp. and *Uromyces* spp. the spines are situated in small, circular depressions in the spore surface, each surrounded by a slightly raised annulus (Figs. 86 and 101) (Amerson and Van Dyke, 1978; Bose and Shaw, 1971; von Brandenburger, 1971; von Brandenburger and Schwinn, 1971; Jones, 1971; Hardwick *et al.*, 1975; Standbridge and Gay, 1969; Rapilly, 1968). In *Puccinia coronata* f. sp. *avenae* the annulus varies from that described above to a low-profile, polygonal configuration of what appears to be only wrinkled pellicle (Corlett, 1970; Takahashi and Furuta, 1973). An even fainter impression of an annulus around the spine base occurs in *Puccinia conoclinii* (Moore and Grand, 1970). Cross sections of some urediospores show a raised area of wall material in approximately the same area as the annulus on the spore surface observed by scanning electron microscopy (Fig. 100) (Amerson and Van Dyke, 1978; Harder, 1976c; Hardwick *et al.*, 1975; Thomas and Isaac, 1967). The significance of the annulus and its association with the development of urediospore spines is discussed later in this Section of Chapter 2.

The cell wall of mature *Melampsora lini* urediospores is approximately 1.0–1.5 μm thick and consists of two or three layers in addition to an extremely thin (7–20 nm) wrinkled pellicle which covers the entire spore, including the spines (Fig. 87) (Littlefield and Bracker, 1971b). In thin section

---

Fig. 77. A "smooth-spored" mutant (described by light microscopy) of *Melampsora lini* has low verrucae (V), approximately 0.1–0.2 μm tall, on its surface when examined by scanning electron microscopy. (× 9,000) (From Littlefield and Bracker, 1971b. Reproduced by permission of the National Research Council of Canada from *Can. J. Bot.* **49,** 2067–2073.)

Fig. 78. The reniform urediospores of *Hemileia vastatrix* are echinulate on their convex (upper) and lateral surfaces, but smooth on their concave (lower) surface. (× 2,500) (L. J. Littlefield, unpublished.)

Fig. 79. Aecioid urediospores of *Chrysomyxa pirolata* are verrucose, a type of ornamentation more typical of aeciospores than urediospores. (× 1,770) (L. J. Littlefield, unpublished.)

Fig. 80. The verrucae (V) of urediospores of *Puccinia colata* are restricted primarily to the ends of the spores, with the central regions being smooth. (× 1,800) (Courtesy of J. F. Hennen, unpublished.)

Fig. 81. *Uredinopsis osmundae* urediospores have one mucronate end (arrow) from which run two ridges along opposite sides of the spore. The second ridge on the spore shown is on the back side, away from the plane of view. The ridge is irregularly thickened, often referred to as coglike in light microscopy. (× 1,450) (L. J. Littlefield, unpublished.)

Fig. 82. Urediospores of *Milesia laeviuscula* have a surface that is essentially smooth. (× 1,350) (L. J. Littlefield, unpublished.)

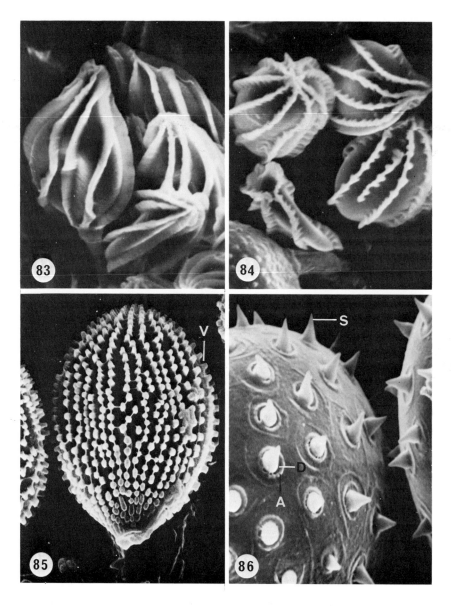

Figs. 83–85. Variations in surface striations of urediospores of *Pileolaria* spp.

Fig. 83. *Pileolaria klugkistiana.* The striations consist of uniformly thin finlike ridges approximately 2.0 μm tall, running the length of the spore, commonly in a partial spiral around the longitudinal axis. (× 1,450) (Courtesy of J. F. Hennen, unpublished.)

Fig. 84. *Pileolaria shiraiana.* The striations have a similar spatial configuration to those of *P. klugkristiana* (Fig. 83), but consist of small knobs connected by thin finlike striations to form continuous rows. (× 1,150) (Courtesy of J. F. Hennen, unpublished.)

of *Puccinia* spp. three wall layers are suggested by variations in electron density (Amerson and Van Dyke, 1978; Sussman *et al.*, 1969). In freeze-fracture preparations of *Melampsora lini* three distinct layers were evident (Fig. 87). Using successive HCl and NaOH digestion, Ehrlich and Ehrlich (1969) demonstrated in *Puccinia graminis* f. sp. *tritici* two wall layers plus the pellicle. This treatment completely removed the outer wall layer, leaving an alveolate inner layer which comprised about 70% of the total thickness of the wall (Fig. 88) and the thin pellicle. The HCl and NaOH treatment completely digested the germ pore region of the spore cell wall, but left the pellicle intact (Fig. 89).

The bulk of the wall of mature urediospores is a secondary layer(s) deposited centripetally to the original primary wall during spore development. Simultaneous with deposition of the secondary wall, the primary wall is digested, and is totally absent in the mature urediospore (Amerson and Van Dyke, 1978).

The thickening of the urediospore wall occurs simultaneous with, and is an essential component of, the development of echinulate ornamentation of urediospores. The following summary of spine development is taken from Amerson and Van Dyke (1978), Harder (1976c), Littlefield and Bracker (1971b), Müller *et al.* (1974a), Takahashi and Furuta (1975a), and Thomas and Isaac (1967). Amerson and Van Dyke (1978), studying *Puccinia sparganioides,* were the first to examine the entire ontogenic process with both transmission and scanning electron microscopy. They provided a comprehensive working model to explain the processes of spine enlargement and displacement and the accompanying increase in urediospore wall thickness (Figs. 90–113). Spines are initiated as small electron-lucent areas immediately outside the plasma membrane at the inner face of the thin wall of the immature urediospore (Fig. 90, inset). Often, the endoplasmic reticulum aggregates in the cytoplasm near the base of the spine initials (Fig. 90). The spines increase in size, and the thickening secondary spore wall layer forms an invaginated collar around the base of the enlarging spine, separating the latter from the plasma membrane (Figs. 92 and 94). As centripetal deposition of secondary wall material continues and as the spines progressively assume a more external position, the inner surface of the wall becomes flat, no longer forming collars around the bases of the spines (Figs. 96, 98, 100).

Fig. 85. *Pileolaria patzcuarensis.* Continuous ridges are absent from the spore surface. Small knoblike verrucae (V), approximately 0.7–1.0μm tall, are arranged in more or less straight rows along the length of the spore, coming to a focus at each end of the spore. ($\times$ 2,300) (L. J. Littlefield, unpublished.)

Fig. 86. Spines (S) on urediospores of *Uromyces phaseoli* var. *typica* reside in small circular depressions (D) encircled by raised annuli (A). ($\times$ 6,100) (Courtesy of J. R. Venette, unpublished.)

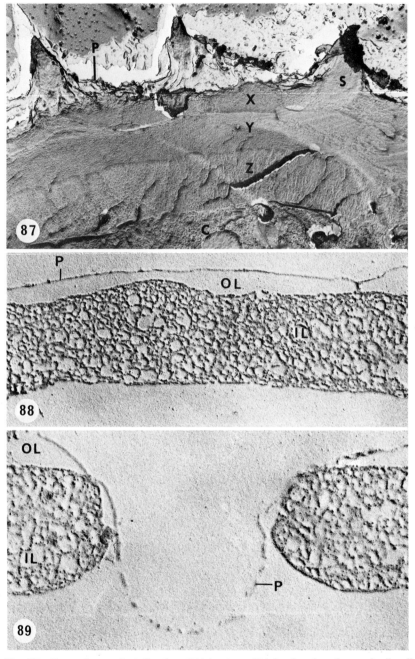

Fig. 87. Freeze-fractured urediospore of *Melampsora lini* showing three layers of cell wall material (X, Y, and Z), the spines (S) partially embedded in the outermost wall layer, and the wrinkled pellicle (P) over the surface of the spore. C, cytoplasm. (× 21,000) (From Littlefield and Bracker, 1971b. Reproduced by permission of the National Research Council of Canada from *Can. J. Bot.* **49,** 2067–2073.)

The spines eventually reside on the spore surface in small circular depressions (Figs. 100 and 101). The seemingly centrifugal displacement of spines is brought about by several simultaneous processes, which are (a) the enlargement of spines from their original small size, approximately 0.1 μm long, to their mature size, approximately 1–1.25 μm long; (b) the dissolution of the primary wall in which they were initially embedded, i.e., in the immature spore; (c) the deposition of secondary wall materials centripetal to the primary cell wall, with the latter eventually disappearing; and (d) the stretching of the secondary wall to assume the full size of mature urediospores. The expansion of the maturing urediospores is evinced by an increased distance between spines and a two- to three-fold increase in spore size when immature and mature spores are compared. The digestion of the primary wall is initiated near the apex of the spines (Figs. 92 and 94) and expands radially until all the primary wall layer disappears. At intermediate stages of development, when primary wall digestion is incomplete, the partially exposed spines are surrounded by a reticulate network comprised of yet undissolved primary wall (Figs. 97 and 99). This process is summarized diagrammatically in Figs. 102–113.

Minor variations from the above-mentioned scheme exist in *Melampsora lini* (Littlefield and Bracker, 1971b) which are (a) no localized digestion of primary wall occurs near the tips of the immature spines; (b) spine maturation is synchronous, whereas it is basipetal in *Puccinia sparganioides* (see Figs. 13 and 14 of Amerson and Van Dyke, 1978); (c) the basal one-fifth to one-quarter, approximately, of spines remain embedded in the spore wall of *Melampsora lini*, whereas the entire spines are superficial in *Puccinia sparganioides* and other examples of *Puccinia* spp. in the literature. Considering the basic similarities in spine development that have been revealed by numerous investigators it appears likely that the scheme proposed by Amerson and Van Dyke (1978) will have broad application in the rust fungi. However, detailed studies of many more genera will be required to identify a common pattern.

The origin of the pellicle which covers the urediospores and their spines is uncertain. In *Melampsora lini* the continuity of striated components of the

---

Figs. 88 and 89.   Walls of urediospores of *Puccinia graminis* f. sp. *tritici* extracted with HCl and NaOH, and shadowed with chromium subsequent to sectioning and removal of the methracylate. (From Ehrlich and Ehrlich, 1969).

Fig. 88.   The outer pellicle (P) and a thick inner wall layer (L) remain after extraction. The region designated (OL) is the void left between the inner wall layer (IL) and the pellicle (P) after HCl and NaOH digestion occupied by the outer wall layer(s) prior to digestion. ($\times$ 17,000)

Fig. 89.   In the germ-pore region of the spore, all wall layers are digested by the treatment, leaving only the pellicle (P) continuous over the pore. IL, inner wall layer; OL, void formed upon extraction of outer layer(s) of spore wall. ($\times$ 21,000)

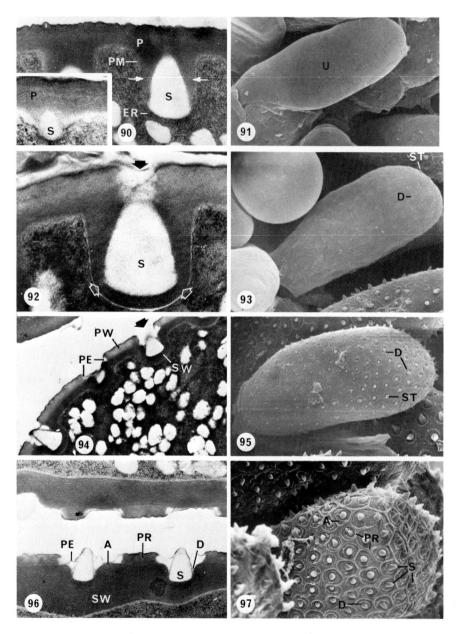

Figs. 90–97 (continued in Figs. 98–101).   Transmission and scanning electron microscopy (TEM and SEM, respectively) of spine development in urediospores of *Puccinia sparganioides*. The stage numbers assigned to Figs. 90–101 correspond to the stages in the diagrammatic summary, Figs. 102–113. [From Amerson and Van Dyke, 1978. Reproduced with permission from *Exp. Mycol.* **2,** 41–50. Copyright Academic Press, Inc. (New York).]

pellicle and intercellular debris of similar composition (Fig. 76) suggest that the pellicle may be a deposition product formed during expansion of the uredium (Hassan and Littlefield, 1979). Intercalary cells and senescent pedicels are crushed and/or autolyzed during continued development of the uredium. These cells contain characteristic lamellar inclusions (Figs. 72–74). Such inclusions in these ephemeral cells, upon crushing and autolysis of the latter, may provide the substance(s) from which the pellicle is derived. Further cytochemical studies are necessary to confirm this hypothesis.

C. Urediospore Germ Pores

The germ pores of urediospores are variable. In *Uromyces phaseoli* var. *typica* (Hardwick *et al.*, 1975) and *Phragmidium tuberculatum* (Henderson and Prentice, 1973) the pore constitutes a region of the spore wall thickened centripetally. However, in both cases most of the locally increased thickness is occupied by a mottled, electron-lucent material interspersed within the wall.

Fig. 90.   Stage 1 (TEM) of urediospore spine development. Note conical spine (S) developing beneath the primary cell wall (P), surrounded by invaginations of that wall (arrows). ER, endoplasmic reticulum; PM, plasma membrane. (× 24,563) Insert: An earlier stage of spine development in which the primary cell wall (PW) invagination has just begun to surround the spine (S) initial. (× 37,306)

Fig. 91.   Stage 1 (SEM) of spine development. Note the smooth surface of the urediospore (U). (× 2,850)

Fig. 92.   Stage 2 (TEM) of spine development showing initial alteration of spore surface. Note encasement of spine (S) with wall secondary material (open arrows) and the apparent erosion of the wall near the apex of the spine (solid arrow). Note also the fibrillar nature of the spine. (× 32,340)

Fig. 93.   Stage 2 (SEM); note the development of small circular depressions (D) containing spine tips (ST), barely visible. (× 2,850)

Fig. 94.   Stage 3 (TEM); note the enlarged eroded areas (arrow) in the primary wall near the spine apex, the secondary wall (SW) which invaginates around the spines, and the pellicle (PE). (× 8,470)

Fig. 95.   Stage 3 (SEM); note the enlarged circular depressions (arrow) and the prominent spine tips (ST). (× 2,250)

Fig. 96.   Stage 4 (TEM); note the progressive displacement of the spines (S) toward the spore surface. Note also the formation of polygonal ridges (PR), the slightly undulating secondary wall (SW), and the formation of depressions (D) and annuli (A) around the spine bases. Separation of the pellicle (PE) from the underlying wall may have occurred during specimen preparation. Stain precipitates are visible in the cell wall. (× 14,380)

Fig. 97.   Stage 4 (SEM); note annuli (A), depressions (D) and spines (S). (× 3,850)

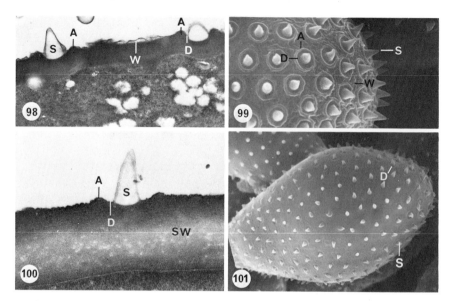

Figs. 98–101. Continuation from Figs. 90–97 showing spine development in urediospores of *Puccinia sparganioides.* [Amerson and Van Dyke, 1978. Reproduced with permission from *Exp. Mycol.* **2,** 41–50. Copyright Academic Press, Inc. (New York).]

Fig. 98. Stage 5 (TEM); note continued spine (S) evagination, and prominent annuli (A) and depressions (D) around spines (S), also the reduction of polygonal ridges (see Fig. 96) to wrinkles (W). (×9,240)

Fig. 99. Stage 5 (SEM); note the prominent annuli (A) and depressions (D), and the absence of polygonal ridges (compared to Fig. 97) which have been reduced to wrinkles (W) on the surface between the annuli. (×4,620)

Fig. 100. Stage 6 (TEM); mature spore. Note that the inner surface of the spore wall (SW) is smooth, not undulate; the spine (S) is fully extended and the annuli (A) and depressions (D) are less prominent than those in previous stages. (×15,400)

Fig. 101. Stage 6 (SEM); mature spore. The spines (S) reside in small depressions (D) surrounded by low annuli. (×1,540)

The germ pores of *Melampsora lini* (Littlefield and Bracker, 1971b; Manocha and Shaw, 1967) and *Puccinia graminis* f. sp. *tritici* (Sussman *et al.,* 1969) are locally thin, with the innermost wall layer being absent or greatly reduced. The interspersion of localized electron-lucent materials in the inner, thin wall layer of the germ pore region was shown in *P. graminis* f. sp. *tritici* (Hess *et al.,* 1975) and throughout nearly the entire thickness of the germ pore in *Uromyces phaseoli* var. *typica* (Macko *et al.,* 1976). Upon acid and base digestion, the entire thickness of the germ pores except the thin, outer pellicle is dissolved from urediospores of *P. graminis* f. sp. *tritici;* the germ pore must therefore differ chemically from the alveolate substance of the inner layer of the spore which remains after such treatment (Fig. 89) (Ehrlich and Ehrlich, 1969).

D.   Urediospore Germination and Host Penetration

The germination process and differentiation of infection structures are discussed in detail in Chapter 3.

Upon germination, urediospores form a germ tube which sometimes branches. Typically the germ tube forms an apical appressorium (Fig. 114) which adheres to the host surface and from which the infection peg enters the host. With few exceptions, urediospore germlings enter the host through stomata, although some, e.g., *Ravenelia humphreyana* and *Puccinia psidii* (Hunt, 1968) and *Phakopsora pachyrhizi* (Marchetti *et al.*, 1975) penetrate directly through the epidermis, as recorded by light microscopy.

VI.   TELIA AND TELIOSPORES

Telia are sori which contain teliospores, the basidia-producing spores of the life cycle (Y. Hiratsuka, 1973). Telia commonly form on the same types of host tissue as do uredia, i.e., leaves, stems, or fronds of herbaceous plants. Generally telia do not form on woody tissue, although *Gymnosporangium* spp. is a conspicuous exception.

The Uredinales are commonly divided into two families, Melampsoraceae and Puccinaceae, based on variations of telial morphology (Arthur, 1934). In either family the teliospores may be sessile or pedicellate, free or laterally united.

In the Melampsoraceae teliospores are typically united into subepidermal crusts, e.g., *Melampsora* spp. (Fig. 115), or hornlike columns, e.g., *Cronartium* spp. (Fig. 116). Such columns extend through the host epidermis and comprise a fascicle of tightly packed, linearly aligned teliospores. Teliospores may also occur singly or grouped within mesophyll cells of the host, e.g., *Uredinopsis* spp., or as loose aggregates within host epidermal cells, e.g., *Milesia* spp. and *Hyalopsora* spp. (Arthur, 1934; Cummins, 1959).

Telia of the Pucciniaceae are commonly erumpent cushions or pulvinate masses that rupture the host epidermis and bear pedicellate teliospores, e.g., *Puccinia xanthii* (Fig. 117). The pedicles of *Diabole* spp. (Fig. 119) are usually dichotomously branched, with each apical cell bearing one teliospore. In *Gymnosporangium* spp. the pedicels are aggregated to form a gelatinous fascicle which bears teliospores at the apices of the individual

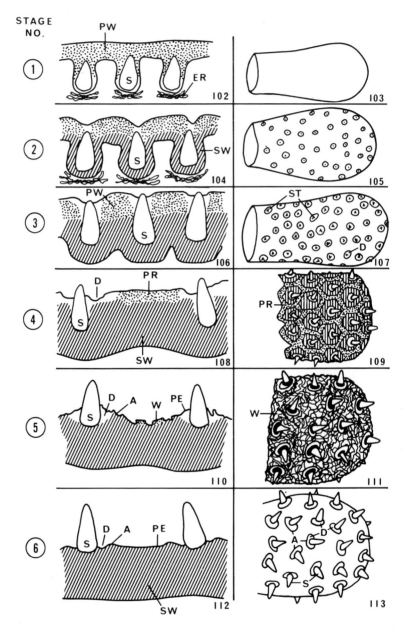

Figs. 102–113. Diagrammatic summary of the ontogeny of urediospore ornamentation of *Puccinia sparganioides*. The six stages of development, the two columns of figures and the figure labels correspond to those of the TEM and SEM micrographs in Figs. 90–101. [From Amerson and Van Dyke, 1978. Reproduced with permission from *Exp. Mycol.* **2**, 41–50. Copyright Academic Press, Inc. (New York).]

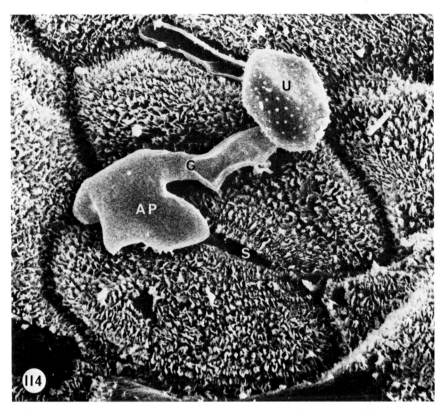

Fig. 114. Urediospores (U) of *Melampsora lini* germinate on host plants under appropriate conditions to produce branched germ tubes (G) that form terminal appressoria (AP) over host stomata (S). An infection peg then penetrates the stoma to initiate infection. (× 2,100) [From Gold and Littlefield, 1979. Reproduced by permission of the National Research Council of Canada from *Can. J. Bot.* (in press).]

Figs. 102 and 103.   Stage 1. Spine development beneath the spore surface.

Figs. 104 and 105.   Stage 2. Spine encasement and initial erosion of wall near spine tips.

Figs. 106 and 107.   Stage 3. Early spine extension.

Figs. 108 and 109.   Stage 4. Mid-spine extension with development of polygonal ridges.

Figs. 110 and 111.   Stage 5. Late spine extension with prominent annuli and depressions.

Figs. 112 and 113.   Stage 6. Mature spore.

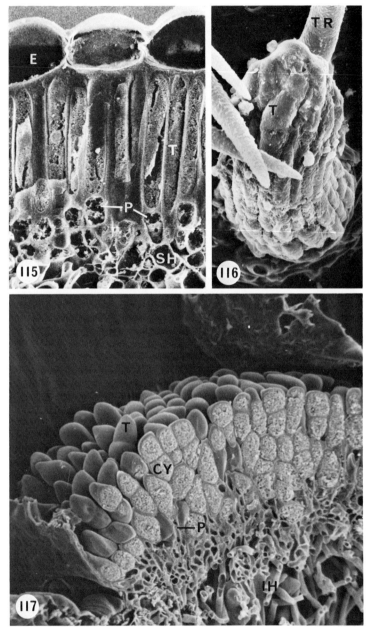

Fig. 115.   Cross fracture of a telial crust of *Melampsora lini*. The prismatic teliospores (T) are borne beneath the host epidermis (E); they arise from sporogenous hyphae (SH) and are borne on short pedicel-like cells (P). (×750) [From Gold and Littlefield, 1979. Reproduced by permission of the National Research Council of Canada from *Can. J. Bot.* (in press).]

pedicels (Fig. 118). Not all species of *Gymnosporangium,* however, e.g., *G. clavipes,* form telial horns, but produce instead pulvinate, cushionlike telia on the surface of fusiform galls (Dodge, 1922). In *Ravenelia* spp. the teliospores are united laterally to form discoid aggregates of spores borne on compound stalks (Figs. 120 and 121). Some species of Ravenelia (Fig. 121) are characterized by small, pendant cysts beneath the compound teliospores (Arthur, 1934; Cummins, 1959). Teliospores of *Prospodium* spp. are borne on pedicels which contain numerous, branched appendages (Fig. 122). Such pedicellate teliospores are borne in distinct cupulate structures often bearing numerous peripheral paraphyses (Fig. 123). The whole telial structure is borne superstomatally (see Hennen and Ono, 1978, for light microscopic comparisons of superstomatal telia). Similarly, teliospores of *Hemileia* spp. are superstomatal, but are borne on short, unbranched pedicels which originate at the apices of inflated basal cells that extend through the host stomata (Cummins, 1959).

Aecioid telia, e.g., *Endocronartium* spp. (Y. Hiratsuka, 1971) and *Kunkelia* spp. (Arthur, 1934), bear teliospores which are morphologically indistinct from most aeciospores. Such teliospores are catenulate, form in association with wedge-shaped intercalary cells, and have verrucose ornamentation (Fig. 124).

A.   Teliospore Formation

Teliospore ontogeny is varied, being related to the particular morphological type to which the telium belongs. Hughes (1970) described three kinds of teliosporogenesis, but only one has received close ultrastructural study. The simplest mode is that found in certain Melampsoraceae where determinate, subepidermal primordial cells give rise individually to only one teliospore. There is no succession of teliospores arising from any one sporogenous cell, e.g., *Pucciniastrum* spp. and *Melampsora* spp. A second method of teliosporogenesis gives rise to a basipetal succession of teliospores that form as meristem arthrospores. In *Didymopsora* spp. the teliospores are separated by intercalary cells (Cummins, 1959) in a manner similar to aeciospores, but in *Cronartium* spp. the intercalary cells are absent. In the Pucciniaceae teliospores develop from successive new growing points on

Fig. 116.   Surface view of a telial column of *Cronartium ribicola,* consisting of cylindrical teliospores (T) aggregated laterally and linearly. TR, host trichome. (× 550) (L. J. Littlefield, unpublished.)

Fig. 117.   Cross fracture of a telium of *Puccinia xanthii.* Note the smooth surface of the two-celled teliospores (T) with their cytoplasm (CY) exposed by the fracture. P, teliospore pedicel; H, intercellular hyphae. (× 500) (From Brown and Brotzman, 1976. Reproduced by permission of the National Research Council of Canada from *Can. J. Microbiol.* **22,** 1252–1257.)

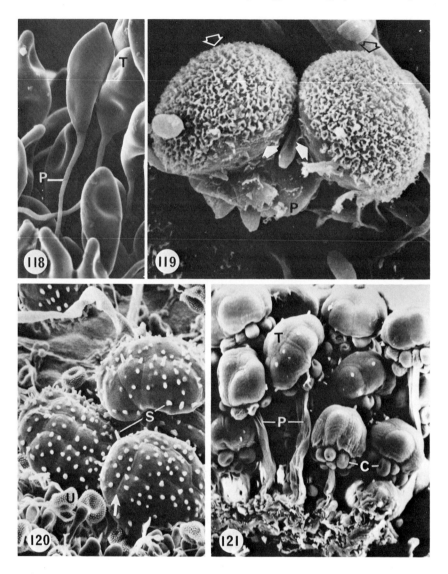

Fig. 118. The smooth surfaced, two-celled teliospores (T) of *Gymnosporangium juniperi-virginianae* are borne on elongated pedicels (P) which swell into gelatinous masses upon wetting. (× 660) (L. J. Littlefield, unpublished.)

Fig. 119. Single-celled teliospores of *Diabole cubensis* are borne on dichotomously branched pedicles (P). The base of this pedicel was broken off below the point of branching prior to or during specimen preparation. The upper surface (open arrows) of the teliospores is ornamented with finely tomentose projections, but the lower surface is smooth (solid arrows). (× 2,700) (L. J. Littlefield, unpublished.)

sporogenous cells, thus being equivalent to sympodioconidia, as are urediospores. Urediosporogenesis and teliosporogenesis in the Pucciniaceae have several features in common. The latter process has been documented ultrastructurally in *Puccinia coronata* f. sp. *avenae* (Harder, 1977) and *Uromyces phaseoli* var. *typica* (Müller *et al.*, 1974b).

According to Harder (1977), teliospores of *Puccinia coronata* f. sp. *avenae* develop in the following manner. Sporogenous cells in the telial primordium bud successively to form spore buds (Fig. 125). The latter then divide into a pedicel cell and the immature teliospore primary cell (Fig. 126). Each of these cells is binucleate. The nuclei of the teliospore primary cell undergo another conjugate division; a septum forms which separates the two pairs of daughter nuclei into a two-celled teliospore (Fig. 127), in contrast to the unicellular urediospores of that genus. The first discernible differences in ontogeny of urediospores and teliospores of this fungus are the early thicken- ing of the cell wall at the apical end of the yet undivided teliospore primary cell (Fig. 126) and the accompanying accumulation of polysaccharide granules in the immature spore. Similar to urediospores, presumed-lipid materials accumulate in the developing teliospores. Ornamentation of these teliospores is discussed later.

The cell walls of these teliospores, except in the region of ornamentation, consist of six distinct layers, $0.55-1.0\mu m$ thick. At a stage late in ontogeny, the cell wall at the base of the spore greatly thickens. Secondary thickening also occurs in the apical wall of the pedicel cell in the septum which separates the apex of the pedicel from the base of the teliospore.

Nuclear fusion of the two haploid nuclei in each cell of the teliospore is ac- companied by apparent disorganization and lack of discernibility of the nuclear envelope. In some cases the nuclear envelopes appear to disintegrate at the point of contact between adjacent nuclei, allowing direct contact of the nucleoplasms.

In *Uromyces phaseoli* var. *typica*, Müller *et al.* (1974b) observed several phenomena similar to the above description of teliosporogenesis in *Puccinia coronata* f. sp. *avenae* (Harder, 1977). The most striking difference, other than the generic difference of single-celled versus two-celled teliospores, is the lack of secondary thickening of the apical wall of the pedicel of *U.*

Fig. 120.   *Ravenelia mimosae-sensitivae.* Discoid heads contain numerous, laterally fused teliospores. The telial aggregates are ornamented with numerous wart to spine-like processes. Arrow, fusion lines between adjacent teliospores; U, urediospores. ($\times 520$) (L. J. Littlefield, un-published.)

Fig. 121.   The fused telial heads (T) of *Ravenelia cassiaecola* are supported by long pedicels (P) and bear sterile, pendant cysts (C) from their lower surface. The surface of the telial ag-gregates is essentially smooth, with only a few verrucae present. ($\times 450$) (L. J. Littlefield, un-published.)

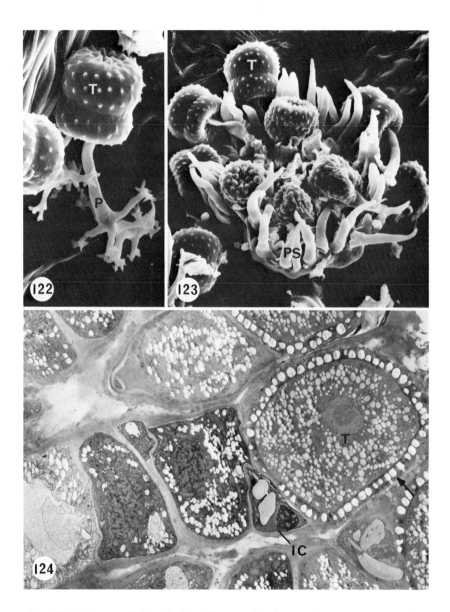

Fig. 122. The two-celled, echinulate teliospores (T) of *Prospodium perornatum* are borne on pedicels (P) which are extensively branched near their base. (×1,150) (L. J. Littlefield, unpublished.)

Fig. 123. The telium of *Prospodium perornatum* is superstomatal, consisting of a basketlike structure bearing the pedicellate teliospores (T) and peripheral, spine-like paraphyses (PS). (×780) (L. J. Littlefield, unpublished.)

Fig. 124. Aecioid teliospores (T) of *Kunkelia nitens* are catenulate, alternating with inter-calary cells (IC) and bear verrucose ornaments (arrow). (×5,000) (Courtesy of C. W. Mims, un-published.)

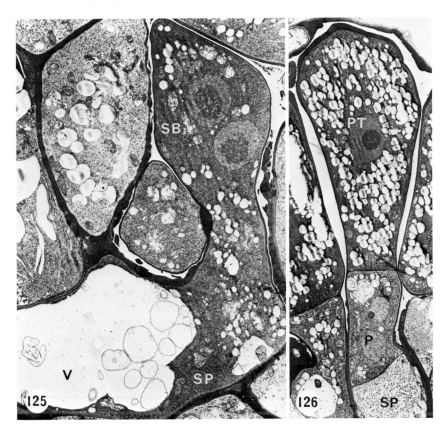

Figs. 125 and 126. Teliosporogenesis in *Puccinia coronata* f. sp. *avenae*. [From Harder, 1977. Reproduced with permission from *Physiol. Plant Pathol.* **10**, 21–28. Copyright Academic Press, Inc. (London), Ltd.]

Fig. 125. A sporogenous cell (SP) has elongated to form a spore bud (SB). V, vacuole. (×6250)

Fig. 126. A later stage in which the teliospore initial has divided to form a primary teliospore (PT) and a pedicel (P). The spore wall has begun to thicken at the apical end. SP, sporogenous cell. (×3,500)

phaseoli var. *typica* compared to *P. coronata* f. sp. *avenae*. A thin apical wall of the pedicel also typifies *Gymnosporangium juniperi-virginianae* (Mims, 1975). With the latter fungus Mims (1975) demonstrated the budding of the sporogenous cell followed by successive development of septa to result in formation of the teliospore initial and the pedicel, similar to the above description of teliosporogenesis in *Puccinia coronata* f. sp. *avenae*.

Ontogeny of aecioid teliospores is the same as that of aeciospores, as shown by the presence of catenulate teliospores having knoblike ornamentation and

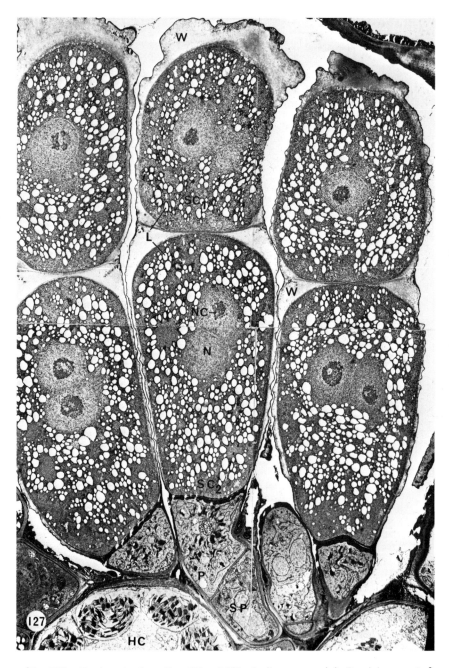

Fig. 127. Continuation from Figs. 125 and 126, of teliosporogenesis in *Puccinia coronata* f. sp. *avenae*; section through a moderately mature telial sorus. The primary teliospore cells (Fig. 126) have each divided to form two-celled, binucleate, teliospores (spore cells 1 and 2 are labelled SC_1 and SC_2 respectively). The spore walls (W) are lightly stained, and wall thickening has continued at the teliospore apex. HC, host cell; L, lipid droplets; N, nucleus; NU, nucleolus; P, pedicel; SP, sporogenous cell. ($\times 3,230$) (From Harder, 1977.)

being separated by wedge-shaped, intercalary cells in *Kunkelia nitens* (Fig. 124) (C. W. Mims, personal communication).

The septal pore in the cross wall separating the mature teliospore from the pedicel cell in *Gymnosporangium juniperi-virginianae* is plugged by electron-opaque material (see Fig. 15 of Mims *et al.*, 1975).

B.   Teliospore Ornamentation

Teliospores generally exhibit a greater variety of ornamentation types than do aeciospores or urediospores. An attempt to categorize the ornamentation types of teliospores as revealed by scanning electron microscopy is a difficult task and, no doubt, different persons would erect some different categories. Ornaments in essentially all rust genera have been described by light microscopy, but the terms so applied do not always conform to the images obtained with the greater resolving power of the electron microscope. Also, arbitrary quantitative distinctions must be made, e.g., the limit between echinulate and spinose, which may vary among investigators. These problems notwithstanding, the following list of teliospore ornamentation types and examples that have been shown ultrastructurally is presented. Terms used to describe the types of ornaments are based on the definitions of Murrill (1905), from light microscopy.

Smooth (no ornamentation on spore surface):
   Cronartium ribicola (Fig. 116, L. J. Littlefield, unpublished)
   Gymnosporangium juniperi-virginianae (Mims *et al.*, 1975) (Fig. 118,
      L. J. Littlefield, unpublished)
   Puccinia bupleuri (Durrieu, 1974)
   Puccinia graminis f. sp. *tritici* (L. J. Littlefield, unpublished)
   Puccinia jackyana (von Brandenburger, 1971)
   Puccinia malvacearum (L. J. Littlefield, unpublished)
   Puccinia saniculae (Durrieu, 1974)
   Puccinia xanthii (Fig. 117, Brown and Brotzman, 1976)
   Uromyces phaseoli var. *typica* (Müller *et al.*, 1974)
   Uromyces fabae (von Brandenburger and Schwinn, 1971)

Smooth (generally, but with a few small projections of various sizes and
      shapes):
   Phragmidium speciosum (Fig. 128, L. J. Littlefield, unpublished)
   Ravenelia cassiaecola (Fig. 121, L. J. Littlefield, unpublished)

Echinulate (spore surface covered with spines):
   Dasyspora gregaria (Figs. 129 and 131, L. J. Littlefield, unpublished)
   Puccinia prostii (Henderson, 1969)
   Prospodium perornatum (Figs. 122 and 123, L. J. Littlefield,
      unpublished)

Figs. 128–134. Variations in teliospore ornamentation.

Tranzschelia sp. (Fig. 130, J. F. Hennen, personal communication)

Tranzschelia fusca (von Brandenburger and Schwinn, 1971)

Verrucose (spore surface covered with warts of varying size, texture and distribution):

Endocronartium harknessii (aecioid teliospores) (Y. Hiratsuka, 1971)

Kunkelia nitens (aecioid teliospores) (Fig. 124, C. W. Mims, personal communication)

Mainsia variabilis (ornaments restricted to apical region of spores) (J. F. Hennen, personal communication)

Phragmidium imitans (Fig. 133, L. J. Littlefield, unpublished)

Phragmidium occidentale (Moore and Grand, 1970)

Puccinia prenanthis (von Brandenburger and Schwinn, 1971)

Puccinia pulvinata (von Brandenburger and Steiner, 1971)

Puccinia smyrnii (von Brandenburger, 1971)

Trachyspora intrusa (Fig. 134, L. J. Littlefield, unpublished)

Uromyces anthyllides (von Brandenburger, 1971)

Uromyces bonaveriae (von Brandenburger, 1971)

Uromyces excavatus (von Brandenburger and Schwinn, 1971)

Uromyces guerkeanus (von Brandenburger, 1971; von Brandenburger and Schwinn, 1971)

Fig. 128.   Teliospores of *Phragmidium speciosum* are smooth except for occasional verrucae (V) which appear to be loosely attached to the spore surface. (× 740) (L. J. Littlefield, unpublished.)

Fig. 129.   Echinulate ornamentation of the two-celled teliospores of *Dasyspora gregaria*. The spines (S) are individually distinct, conical structures, approximately 1.5–1.7 µm tall. (See Fig. 131 also.) P, pedicel. (× 1,460) (L. J. Littlefield, unpublished.)

Fig. 130.   Echinulate ornamentation of *Tranzschelia* sp. teliospores. The coarse spines (S) are interconnected by thin ridges (R) of presumed cell-wall material. (× 1,800) (Courtesy of J. F. Hennen, unpublished.)

Fig. 131.   Detail of echinulate ornaments of *Dasypora gregaria* teliospores; see Fig. 129 also. The spines near the ends of the spore are often branched at their apices (open arrow), but the remainder are conical (solid arrow). (× 6,200) (L. J. Littlefield, unpublished.)

Fig. 132.   The spinose projections (S) on the telial surface of *Ravenelia stevensii* are typically branched at their apices. F, line denoting plane of fusion between adjacent teliospores in the compound telial head. (× 2,600) (Courtesy of R. L. Gilbertson, unpublished.)

Fig. 133.   The entire surface of *Phragmidium imitans* teliospores is covered with closely spaced, irregularly shaped verrucae (V). (× 1,100) (L. J. Littlefield, unpublished.)

Fig. 134.   The surface verrucae (V) of *Trachyspora intrusa* teliospores are randomly distributed among more or less smooth-surfaced areas. (× 780) (L. J. Littlefield, unpublished.)

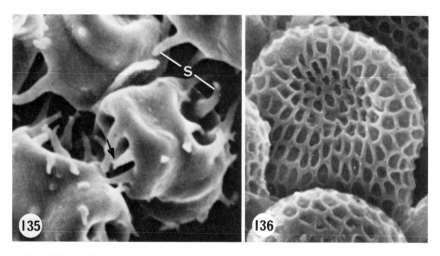

Figs. 135 and 136.   Variations of teliospore ornamentation, continued.

Fig. 135.   Teliospores of *Nyssopsora* sp. are three-celled and bear large, tapering spines (S) 6–13 μm long, some of which are minutely branched at the apex (arrow). (× 1,100) (Courtesy of J. F. Hennen, unpublished.)

Fig. 136.   Reticulate ornamentation of teliospores of *Puccinia echinopteridis*. (× 1,350) (Courtesy of J. F. Hennen, unpublished.)

Spinose (spore surface bears elongated spines of various lengths and degrees of branching):
>   *Nyssopsora* sp. (Fig. 135, J. F. Hennen, personal communication)
>   *Nyssopsora echinata* (Henderson, 1969)
>   *Ravenelia stevensii* (Fig. 132, R. L. Gilbertson, personal communication)

Apically digitate (apex of spore bears thick, digitate projections):
>   *Puccinia coronata* f. sp. *avenae* (Fig. 137, Harder, 1976c; Takahashi and Furuta, 1973) (Fig. 138, R. E. Gold, personal communication)

Reticulate (spore surface netlike, covered with more or less rectangular depressions separated by ridges):
>   *Puccinia echinopteridis* (Fig. 136, J. F. Hennen, personal communication)
>   *Puccinia retifera* (Durrieu, 1974)
>   *Uromyces erythronii* (von Brandenburger and Schwinn, 1971)

Striate (spore surface covered with linear depressions separated by more or less parallel ridges):
>   *Cumminsiella wootoniana* (Fig. 139, J. F. Hennen, personal communication)

Puccinia laurifoliae (Fig. 140, J. F. Hennen, personal communication)
Puccinia rugulosa (Durrieu, 1974)

Rugose (spore surface irregularly sculptured, having a rough, wrinkled appearance, lacking regularly spaced or uniformly sized depressions and projections):

Pileolaria brevipes (J. F. Hennen, personal communication)
Pileolaria standleyi (J. F. Hennen, personal communication)
Pileolaria toxicodendri (Fig. 142, L. J. Littlefield, unpublished)
Puccinia eremuri (von Brandenburger and Steiner, 1971)
Puccinia nitida (Durrieu, 1974)
Puccinia prionosciadii (J. F. Hennen, personal communication)
Uromyces hedysari-paniculati (Moore and Grand, 1970)

Punctate (spore surface sharply pitted with numerous, sometimes irregularly shaped depressions; total area of depressed surface generally less than the raised surface):

Puccinia cervariae (Durrieu, 1974)
Puccinia chaerophylli (Durrieu, 1974)
Puccinia eryngii (Durrieu, 1974)
Puccinia pimpinellae (Fig. 141, L. J. Littlefield, unpublished; von Brandenburger and Schwinn, 1971; Durrieu, 1974)

Tomentose (spore surface covered with finely matted hairs, having a somewhat feltlike appearance):

Diabole cubensis (Fig. 119, L. J. Littlefield, unpublished; tomentose texture restricted to upper surface of spore; lower surface smooth)

Doubtless, as scanning electron microscopy is extended to rust species not yet examined, additional types and variations of ornamentation will be revealed.

The ultrastructure and ontogeny of teliospore ornamentation have been studied in a limited number of cases. The apical digitate projections in *Puccinia coronata* f. sp. *avenae* form as extensions of the thickened spore cap, thus forming the "crown" for which the species is named. Longitudinal sections (Fig. 137) show that the digitate projections consist of several layers of wall material similar in appearance to cell-wall material in the remainder of the spore (Harder, 1977). Studies were not made to show the manner in which this layered wall material is deposited. Thin sections of the spines on echinulate teliospores of *Puccinia prostii* show a uniform wall with the great mass of the spine only slightly different in appearance from the thin, slightly more electron-opaque outer layer (Henderson, 1969).

Henderson *et al.* (1972a) proposed that formation of surface topography in punctate teliospores of *Puccinia chaerophylli* results from localized enzymatic activity within the cell wall. At an early stage of development, small "dense

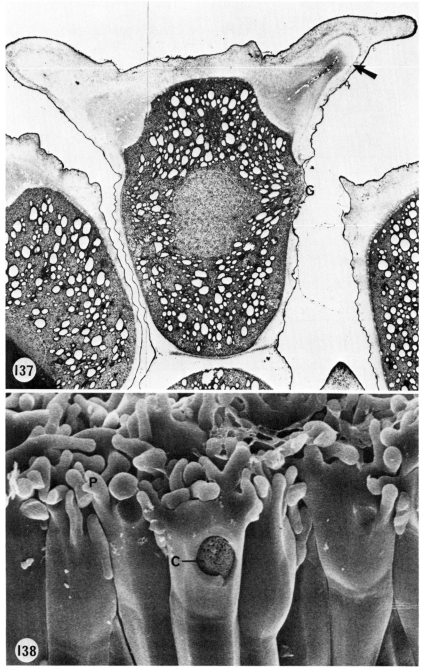

Figs. 137 and 138.   Teliospores of *Puccinia coronata* f. sp. *avenae*.

granules" are present in the inner layer of the cell wall near the plasma membrane. Gradually they are displaced centrifugally within the wall; simultaneously, electron-lucent halos form around the granules. Henderson *et al.* (1972a) suggested that the granules are centers of enzymatic activity, leading to digestion and shrinkage of parts of the wall, which produce localized depressions separated by ridges that are not so affected. Lateral migration and coalescence of granules, or the disappearance of some of them, results in more or less uniformly concentrated aggregations of granules to provide for the spaced depressions in the spore wall. Durrieu (1974) discussed possible phylogenetic relationships of variously ornamented forms of teliospores and their association to the postulated enzymatic activity. Germ pore development is similar to that of the individual depressions, but the enzymatic activity apparently is more pronounced there as the germ pore appears to have undergone more digestion and shrinkage than other parts of the spore wall (Durrieu, 1974; Henderson *et al.*, 1972a).

C. Teliospore Germ Pores

The structure of teliospore germ pores is variable. The apical germ pore of *Uromyces phaseoli* var. *typica* teliospores consists of a locally thickened region of cell wall containing electron-lucent areas in which appear to be electron-opaque granules and vesicular bodies (Müller *et al.*, 1974b). This contrasts to the germ pores of *Gymnosporangium juniperi-virginianae* teliospores in which the pores are locally thin regions of the wall surrounded by an annulus of thickened wall (Mims *et al.*, 1975). The teliospore wall of the latter contains several layers. No partially disintegrated appearance or irregular intermixing of electron-lucent and opaque areas characterize germ pores in this fungus. In surface view the germ pore in *Puccinia chaerophylli* is a locally depressed zone with an irregular margin and rough surface (Durrieu, 1974).

D. Teliospore Germination and Basidiospore Formation

Teliospores and associated structures have been the subject of numerous terminological disputes for many years, as summarized by Petersen (1974).

Fig. 137. Longitudinal section through the upper cell of a two-celled teliospore showing the extensive apical wall thickening and the digitate processes which extend from the apex. Note the marked reduction in cell-wall thickness in the region of the germ pore (G), also the multilayered wall (arrow) within the digitate ornaments. (× 3,900) [From Harder, 1977. Reproduced with permission from *Physiol. Plant Pathol.* 10, 21–28. Copyright Academic Press, Inc. (London), Ltd.]

Fig. 138. Surface view showing the apically digitate processes (P). Note cytoplasm (C) exposed by a tangential cut through the cell wall. (× 1,700) (Courtesy of R. E. Gold, unpublished.)

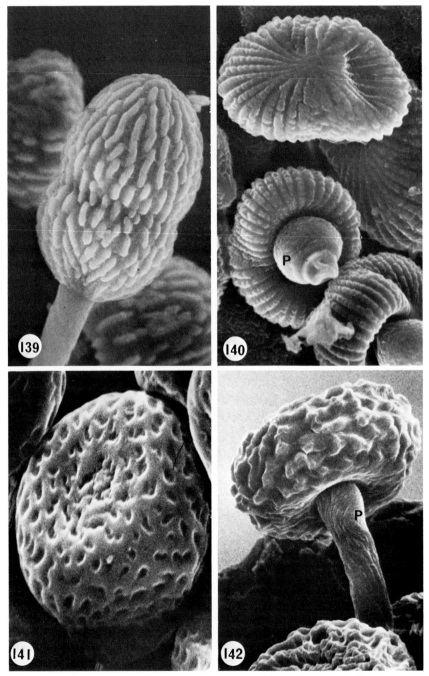

Figs. 139–142.   Variations of teliospore ornamentation, continued.

For our discussion, which is devoted entirely to descriptive morphology without evolutionary considerations, we follow the terminology of Martin (1957) and Ainsworth (1971). The teliospore is the probasidium, that portion of the basidium in which karyogamy occurs. Following karyogamy, meiosis proceeds within the metabasidium. Thus, the teliospore, the probasidium, and the metabasidium are all the same structure in those rusts whose teliospores undergo "internal germination," e.g., *Coleosporium* spp., see below. Because of the possible confusion resulting from a terminology in which a "teliospore," upon germination, produces an "internal pro-mycelium" (i.e., a metabasidium), Olive (1948) preferred to use the term "basidial sorus" instead of "telium" when describing *Coleosporium ver-noniae*. In rusts that form an external metabasidium, e.g., *Puccinia* spp., that structure is distinct from the probasidium, or teliospore. The metabasidium, whether internal or external, typically divides into four cells, each of which will produce a basidiospore on a sterigma. Exceptions to this are reviewed by Petersen (1974). The metabasidium was commonly termed "promycelium" and the basidiospores "sporidia" in the older literature.

Internal germination of teliospores, i.e., the transformation of the teliospore (= probasidium) into a metabasidium which gives rise to four elongated sterigmata-bearing basidiospores, has not been studied ultrastruc-turally.

In the subepidermal teliospores of *Melampsora lini* germination proceeds by the formation of an apical metabasidium which extends beyond the host surface and elongates parallel to that surface (Fig. 143). Following meiosis of the diploid nucleus within the metabasidium, septa form, segmenting the structure into four cells. Each of the four metabasidial cells gives rise to a conical sterigma approximately 4.0 μm tall which bears a basidiospore at its apex (Figs. 1 and 144). Seldom are all four basidiospores observed simultaneously as they commonly develop and are discharged sequentially (Gold and Littlefield, 1979).

Fig. 139. Surface striations on teliospores of *Cumminsiella wootoniana* consist of variable-length ridges of wall material oriented longitudinally. ($\times$ 2,400) (Courtesy of J. F. Hennen, un-published.)

Fig. 140. Surface striations of teliospores of *Puccinia laurifoliae* radiate from the center of the upper surface, around the sides, and to the lower surface of the spore. Note the inflated pedicel (P) cell. ($\times$ 1,200) (Courtesy of J. F. Hennen, unpublished.)

Fig. 141. The punctate surface of *Puccinia pimpinella* teliospores is characterized by small, distinct depressions in an otherwise smooth surface. ($\times$ 3,050) (L. J. Littlefield, unpublished.)

Fig. 142. The rugose, irregular wrinkled surface of *Pileolaria toxicodendri* teliospores covers the entire surface, but is markedly reduced in intensity on the lower surface near the point of attachment to the pedicel (P). ($\times$ 2,100) (L. J. Littlefield, unpublished.)

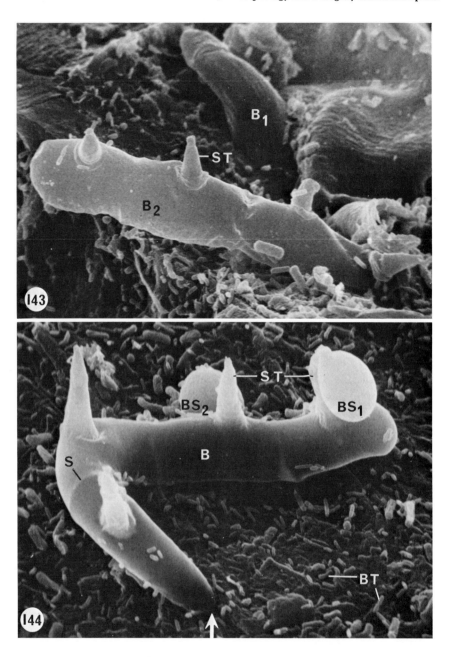

Figs. 143 and 144. Metabasidia of *Melampsora lini*, formed upon germination of teliospores. [From Gold and Littlefield, 1979. Reproduced by permission of the National Research Council of Canada from *Can. J. Bot.* (in press).]

Similar structures of the metabasidium, sterigmata and basidiospores occur in *Gymnosporangium juniperi-virginianae* (Mims, 1977a) and *Cronartium ribicola* (Figs. 145 and 146) (L. J. Littlefield, unpublished). Scanning electron microscopy of *C. ribicola* (Fig. 146) fails to confirm the presence of a socketlike structure (the "caviculum") at the apex of the sterigma in which the base of the basidiospore has been suggested to be attached (Bega and Scott, 1966).

Germ tubes often emerge from both germ pores of the teliospores of *Gymnosporangium* spp. (Mims *et al.*, 1975; Takahashi and Furuta, 1975b), but one becomes dominant and develops into the metabasidium. The cell wall of the emerging metabasidium is clearly continuous with the thin, innermost layer of the teliospore cell wall (Fig. 147) (Kohno *et al.*, 1975; Mims *et al.*, 1975). Germination of *Gymnosporangium* spp. teliospores is accompanied by gelatinization and extensive swelling (up to several millimeters long) of the teliospore pedicels.

Aecioid teliospores of *Endocronartium harknessii* germinate to form elongated germ tubes (homologous with metabasidia) which bear no basidiospores (Y. Hiratsuka, 1969). Such germ tubes apparently penetrate host needles directly through the epidermis (True, 1938), although, based on light microscopic evidence, some uncertainty exists concerning this question (Y. Hiratsuka and Powell, 1976).

VII. OVERVIEW

Different sori and spores formed during the life cycle of rust fungi are characterized by certain morphological and functional properties. Whether one chooses to emphasize the morphological properties (Laundon, 1973) or the functional properties (Y. Hiratsuka, 1973) provides a basis for major controversies among uredinologists. Disregarding those arguments for the moment it is worthwhile to examine some of the structures and processes which typify the respective stages of the life cycle ultrastructurally. Those characteristic features summarized in Table 1 may be of diagnostic value to

---

Fig. 143. Young metabasidia (B_1) penetrate the host epidermis covering the telium, bend approximately 90° and continue growth parallel to the surface. An older metabasidium (B_2) bears four sterigmata (ST) on each of its cells. Basidiospores are absent, having not yet formed, or having been previously ejected. ($\times$ 3,650)

Fig. 144. A mature metabasidium, with three septa (S) which separate the four cells of the structure. A basidiospore (BS_1) is attached to the terminal sterigma (ST). A second basidiospore (BS_2), behind the metabasidium, was perhaps attached previously to the neighboring sterigma (ST). BT, bacteria on the telial surface; arrow, point of origin of the metabasidium from the underlying, vertically oriented teliospore, out of the field of view beneath the telial surface. ($\times$ 3,250)

TABLE 1

Some Morphological Characteristics of Stages of the Life Cycle of Rust Fungi[a]

Characteristic	Basidial stage	Pycnial stage	Aecial stage	Uredial stage	Telial stage
I. Ontogeny of spores and alterations in their cell walls during spore formation	Basidiospores develop presumably in a holoblastic manner from the sterigma apex; no ultrastructural studies yet made on development of the spore wall. Wall of mature spore is one-layer thick except for a second, thin, outer layer restricted to the perihilar region (*Gymnosporangium juniperus-virginianae*).	Pycniospores develop annellidically, but have certain phialidic properties. Mature wall contains two layers; during maturation of wall the thickness remains constant, but the inner (secondary) layer thickens as the outer (primary) layer diminishes.	Aeciospores develop by arthrosporic cleavage from spore initials that form annellidically from sporogenous cells. Mature spore cell wall is single-layered; initially two-layered, but outer (primary) wall layer disappears as the inner (secondary) layer continues to thicken.	Urediospores develop sympodially. Blastic development of spore bud from sporogenous cell; cleavage of the spore bud from the sporogenous cell to form a urediospore initial; arthrosporic cleavage of the urediospore initial into an immature urediospore and a pedicel. Mature cell wall of two to three layers, increasing greatly in thickness during maturation. During maturation the outermost (primary) wall layer gradually erodes and is replaced by the thicker, inner (secondary) layer(s).	Teliospore ontogeny is varied: sympodial (the "Pucciniaceous" mode), meristem arthrosporic (the "Cronartiaceous" mode) or by transformation of a terminal primordial cell, without sympodial proliferation (the "Melampsoraceous" mode) (Hughes, 1970). Up to six discernible cell-wall layers in mature spores; typically thicker wall than other spore types.
II. Spore ornamentation	Basidiospores are typically smooth to slightly rugose; very few species studied ultrastructurally.	Pycniospores are typically smooth; very few species studied ultrastructurally.	Aeciospores are typically verrucose, a few somewhat echinulate. Ornaments initiated and attain full size within primary cell wall. They become exposed as this wall is dissolved and the secondary wall layer is laid down to the inside of the dissolving primary wall.	Urediospores are typically echinulate; some verrucose, ridged or smooth. Ornaments initiated at inner face of primary (inner) cell wall layer; concomitant with their enlargement, the primary wall erodes and is replaced by the thick secondary wall which subtends the ornaments at maturity. (This process is similar to ornamentation ontogeny in aeciospores, but with significant differences; see discussion in this chapter.)	Teliospores have extremely varied ornamentation: apically digitate, echinulate, punctate, reticulate, rugose, smooth, spinose, striate, tomentose. Ontogeny of ornaments not known.

88

III. Accessory structures present	Accessory structures are lacking in this stage.	Flexuous hyphae are presumably present in all pycnia. Ostiolar trichomes, i.e., periphyses and/or paraphyses, may be present (e.g., *Puccinia graminis*) or absent (e.g., *Melampsora lini*).	Intercalary cells occur between catenulate aeciospores, but disintegrate as aeciospores mature. A peridium surrounds the aecium; it is more pronounced in some aecial types, e.g., roestelioid, than others, e.g., caeomoid.	Peridial and intercalary cells and paraphyses are common in Melampsoraceae, but not Pucciniaceae. Inflated basal cells to which pedicels are attached occur in some superstomatal forms, e.g., *Hemileia* spp.	Accessory structures are typically absent except for sterile basal cells (sometimes inflated) and paraphyses in some superstomatal forms, e.g., *Prospodium* spp., or intercalary cells in genera with catenulate teliospores, e.g., *Didymopsora* spp. or sterile, pendant cysts, e.g., some *Ravenelia* spp.
IV. Cell wall continuity upon germination (limited number of ultrastructural observations)	Ultrastructural study limited to *Gymnosporangium juniperi-virginianae*; total thickness of basidiospore wall is continuous with that of the germ tube.	Pycniospores fuse with flexuous hyphae; no "true" germ tube produced. No ultrastructural studies yet made of this fusion process.	Ultrastructural study limited to *Cronartium fusiforme*. There the germ-tube wall appears to form *de novo* within the pore region; it is not continuous with any discernible wall layer in that region.	Germ-tube wall is typically continuous with only the innermost layer of the urediospore wall. That layer exists only in the region of the germ pore, tapering in thickness and eventually disappearing as one moves away from the germ pore.	Ultrastructural observations limited to *Gymnosporangium juniperi-virginianae*. The wall of the metabasidium is continuous with the innermost layer of the teliospore wall. This layer exists throughout the spore, not being restricted to the region of the germ pore.
V. Route of host penetration during infection	Basidiospore germlings typically penetrate directly through cuticle into epidermis; stomatal penetration by *Cronartium ribicola*.	Pycniospore germlings do not proceed to penetrate host; no ultrastructural documentation of host penetration by pycniospore germ tubes of *Melampsora lini* reported by R. F. Allen (1934) using light microscopy.	Aeciospore germlings typically penetrate via stomata, but this recorded only by light microscopy; direct penetration by *Gymnoconia interstitialis* (Pady, 1935).	Urediospore germlings typically penetrate via stomata, but enter directly through cuticle and epidermis in *Puccinia psidii* and *Ravenelia humphreyana* (light microscopy).	Teliospores germinate to form metabasidia; no host penetration.
VI. Types of haustoria produced (discussed in Chapter 4, Section II)	Basidiospores give rise to M-haustoria (i.e., monokaryotic, intracellular, hypha-type morphology).	Pycniospores do not proceed to form haustoria.	Aeciospores infect to give rise to D-haustoria (i.e., dikaryotic, "typical" stalked haustoria with narrow neck and bulbous body); insufficient studies have been made to note exceptions. M-haustoria present in host tissue *containing* the aecial stage are derived from basidiospore infection.	Urediospores infect to give rise to D-haustoria, with the exceptions of M-haustoria in some tropical rusts (Hunt, 1968). D-haustoria present in host tissue *containing* the uredial stage are derived from either urediospore or aeciospore infection.	Teliospores do not proceed to form haustoria. D-haustoria typically present in host tissue *containing* the telial stage are derived from either aeciospore or urediospore infection. The M-haustoria present in host tissue *containing* teliospores (e.g., *Puccinia malvacearum* and other microcyclic rusts) are derived from basidiospore infection.

a See text of this chapter for examples and references when not given, except for Characteristic VI. See Chapter 4, Section II for discussion of that topic.

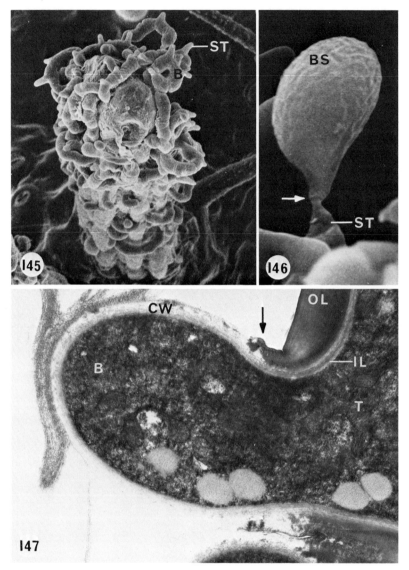

Figs. 145 and 146.   Germination of teliospores of *Cronartium ribicola*. (L. J. Littlefield, un-published.)

Fig. 145.   Numerous metabasidium (B) on the surface of a telial column. The sterigmata (ST) of immature metabasidia are broad and have rounded apices. (× 540)

Fig. 146.   Basidiospore (BS) borne at the apex of a sterigma (ST), the two structures being separated by a septum (arrow). (× 3,520)

Fig. 147.   The outer cell-wall layer (OL) of the teliospore (T) of *Gymnosporangium juniperi-virginianae* is ruptured (arrow) by the emerging metabasidium (B). The cell wall (CW) of the basidium is continuous with the inner wall layer (IL) of the teliospore. (× 42,500) (Courtesy of C. W. Mims, unpublished.)

the taxonomist and may reveal relationships which heretofore have not been recognized. The topics include (I) Ontogeny and alterations in cell walls during spore formation, (II) Spore ornamentation, (III) Accessory structures present, (IV) Cell wall continuity upon spore germination, and (V) Route of host penetration. Although not discussed in this chapter (see Chapter 4, Section II for details), the types of haustoria produced are also useful in characterizing the stages of the life cycle of rust fungi. This information is included as characteristic VI in Table 1.

Some of the characters listed in Table 1 are more widely documented taxonomically than others (see corresponding sections of this chapter for details and examples, or Chapter 4, Section II for characteristic VI). Also, this tabular summary does not take into account the anomalies of uredinoid aecia, aecioid uredia, aecioid telia, etc. (Y. Hiratsuka, 1973). Obviously such generalizations as this are fraught with danger, and they are presented here only as a working hypothesis based on limited ultrastructure knowledge, certainly not as the final word on comparative morphology of the Uredinales. As research continues into the ultrastructure of rust fungi more precision and wider documentation will be possible when constructing such a list; probably some of our statements will be proven invalid.

Extensive gaps exist in our knowledge of the Uredinales. Perhaps our greatest ignorance in the biology of rust fungi is the dikaryotization process, a most fertile area for ultrastructural research. Another essentially untouched topic is that of ornamentation ontogeny in teliospores. The question of haustorial morphology as related to stage of the life cycle (see Chapter 4, Section II) has received considerable interest recently, but only of a descriptive nature. The question "why" as well as the phylogenetic implications of this phenomenon have yet to be addressed. Thus what may appear to the casual observer as a wealth of information about the Uredinales is in fact only a mere beginning.

3

Infection of the
Susceptible Host

Most ultrastructural studies of the process of infection have concentrated on infections derived from the urediospore germ tube; thus this chapter will deal primarily with the initiation of the uredial stage of the rust life cycle. For comparative purposes, certain features of urediospore germination will also be mentioned but other aspects of the germination of these and the remaining types of rust spores are covered in Chapter 2.

I. DIRECTIONAL GROWTH OF UREDIOSPORE GERM TUBES

In contrast to the direct penetration of basidiospore germ tubes into epidermal cells (see Chapter 2, Section I), germ tubes of urediospores usually have to form an appressorium over a stoma before infection can occur (see Fig. 1; see also Fig. 2 discussed in Section III,A). The relative precision with which germ tubes carry out this process has been long recognized (e.g., R. F. Allen, 1928) and appears to be the result of two processes, directional growth towards the stoma and induction of the appressorium once the stoma has been encountered.

That directional growth may be a thigmotropic response to the plant surface was first postulated by Johnson (1934) who noted that urediospore germ tubes of *Puccinia graminis* f. sp. *tritici* grow predominantly along the transverse axis of the plant leaf. This hypothesis was supported by the light microscopic observation that *P. antirrhini* germ tubes grow at right angles to the parallel ridges of the host cuticle, and will do so even on isolated cuticles or leaf replicas (Maheshwari and Hildebrandt, 1967). Subsequently, Dickinson (1969) demonstrated that several *Puccinia* species will grow at right angles (although sometimes in a zigzag fashion) to a repetitive series of

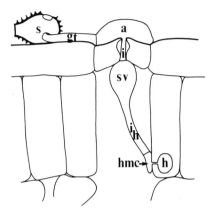

Fig. 1.   Diagrammatic representation of a cross section of a leaf showing the infection struc-tures typically derived from a urediospore (s) on the leaf surface. The germ tube (gt) has sequen-tially formed an appressorium (a) over a stoma, an infection peg (i) between the guard cells, a substomatal vesicle (sv), infection hypha (ih), and haustorial mother cell (hmc) in the in-tracellular space, and a haustorium (h) in a host mesophyll cell. Each rust species has a characteristic infection structure morphology; the one shown here is typical of *Uromyces phaseoli* var. *vignae.*

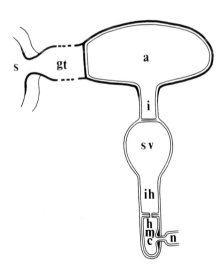

Fig. 2.   Diagram showing the morphological changes in wall structure which are found in urediospore-derived infection structures of *Uromyces phaseoli* var. *vignae;* for details of wall structure of the haustorial mother cell (hmc) and haustorial neck (n) (see Chapter 4, Section II, A,1). For remaining abbreviations, see Fig. 1.

straight structural features in artificial membranes. From the results of a transmission electron microscopic study of replicas of dried, hydrochloric acid-treated germ tubes, he suggests that these structural features of the membrane bring about directional growth by changing the orientation of the adpressed microfibrils of the fungal wall (Dickinson, 1977).

Although cuticular ridges seem to be the most obvious repetitive surface feature of *Antirhinnum majus* surfaces (Maheshwari and Hildebrandt, 1967), scanning electron microscopy has not shown any similar feature for *Phaseolus vulgaris* but has illustrated that germ tubes of *Uromyces phaseoli* var. *typica* grow at right angles to the larger ridges formed by the curvature of the host epidermal cells (Fig. 3) (Wynn, 1976). However, Lewis and Day (1972) point out that since the part of the leaf in direct contact with the germ tube is epicuticular wax, this must be the only part to which the fungus can respond. From stereopair, scanning electron micrographs they conclude that germ tubes of *Puccinia graminis* f. sp. *tritici* orient themselves along one axis of the crystal lattice of this wax layer; the orientation of the lattice relative to the leaf results in the observed transverse growth of the germ tube (Fig. 4). Certainly the adhesion by *P. graminis* f. sp. *tritici* to the host surface seems to depend on the wax being present since germ tubes will not adhere if the wax is removed by organic solvents (Fig. 5) (Wynn and Wilmot, 1977). However, it is difficult to reconcile the hypothesis of the epicuticular wax as being the *only* thigmotropic stimulus with the observation that directional growth of rusts other than *P. graminis* f. sp. *tritici* occurs on leaf replicas where this crystal lattice is missing (Maheshwari and Hildebrandt, 1967; Wynn, 1976). Thus the exact nature of the thigmotropic stimulus on the leaf surface is still an open question and may well be different for different rust species. In all cases, however, the end result of directional growth is that, because of the arrangement of the stomata in relation to the possible thigmotropic stimuli described above, the fungus greatly increases its chances of encountering a stoma and successfully infecting the leaf (Figs. 3 and 4).

II. INDUCTION OF APPRESSORIA

The urediospore germ tube not only has to "find" a stoma but it then has to "recognize" it in some way in order to trigger the formation of the appressorium. Onoe *et al.* (1972) suggest from a scanning electron microscope study of *Puccinia coronata* that this trigger may be the coating of the fungus by wax dissolved from the leaf surface. However, evidence from appressorium induction by artificial membranes (Fig. 6) (Dickinson, 1949, 1970, 1971; Heath, 1977; Maheshwari *et al.*, 1965, 1967) suggests that at least eight species of rusts (including *P. coronata*) can form appressoria in response to a purely thigmotropic stimulus; *P. graminis* f. sp. *tritici* is the

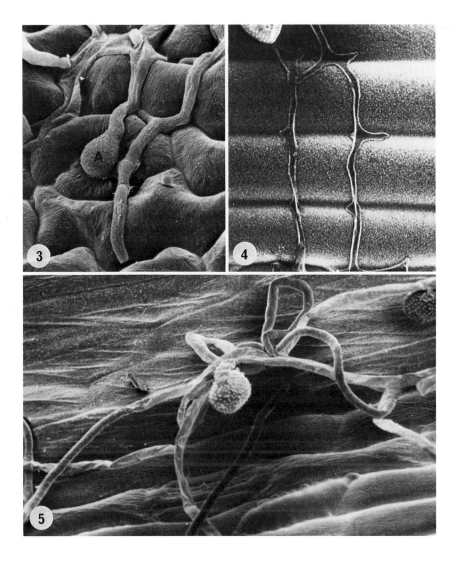

Figs. 3–5. Scanning electron micrographs of the behavior of urediospore germ tubes on host leaf surfaces.

Fig. 3. Appressorium (A) of *Uromyces phaseoli* var. *typica* formed over a leaf stoma of *Phaseolus vulgaris*. An adjacent germ tube has grown over the already occupied stoma without forming an appressorium. (× 750) (Wynn, 1976.)

Fig. 4. Germ tubes of *Puccinia graminis* f. sp. *tritici* growing across the ridges of a normal, wax covered, *Triticum aestivum* leaf. (× 525) (Fixed in osmium tetroxide vapor, courtesy of V. A. Wilmot and W. K. Wynn, unpublished.)

Fig. 5. Germ tubes of *Puccinia graminis* f. sp. *tritici* growing on the surface of a *triticum aestivum* leaf from which the surface wax has been removed by a 5 second dip in chloroform. Note the unoriented growth of the germ tubes compared with Fig. 4, and the lack of adhesion to the leaf surface. (× 670) (Fixed in glutaraldehyde and osmium tetroxide; courtesy of V. A. Wilmot and W. K. Wynn, unpublished.)

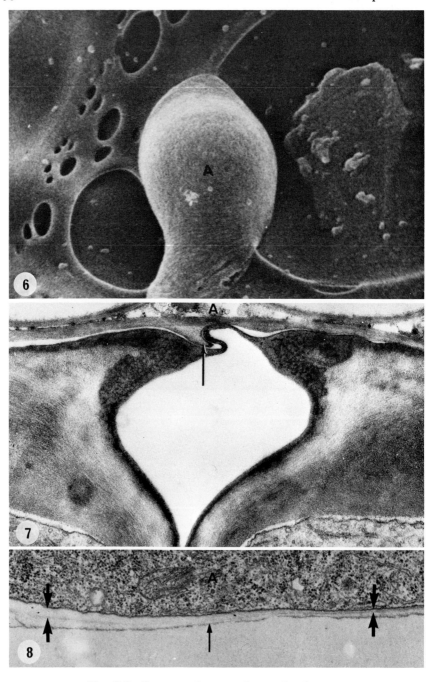

Figs. 6–8.   Responses of appressoria to surface features.

only notable exception where consistent appressorium formation away from the host can be obtained only by temperature shock (Dunkle and Allen, 1971) or by chemical means (Allen, 1957; Grambow and Riedel, 1977; Maheshwari *et al.*, 1967).

Using artificial membranes containing granules of different diameters, Dickinson (1970) found that the minimum diameter capable of inducing appressoria in *Puccinia recondita* is about 3 μm. He calculated that the curved surface of a sphere this size represents an effective unit of stimulus about 120 × 1.2 nm and points out that this is about the size of a cellulose molecule. As yet, however, there is little evidence to suggest the nature of the effective stimulus in nature. Using scanning electron microscopy of artificial membranes, Wynn (1976) found that *Uromyces phaseoli* var. *typica* forms appressoria only over the relatively large ridges formed by the edges of craters and scratches (Fig. 6) and suggests that similar "natural" stimuli on the leaf surface may be the protruding stomatal lips (see Chapter 6, Fig. 2). However, because of the necessity for close adherence of the germ tubes to the surface before appressoria can form (Wynn, 1976), and the way in which these appressoria have been shown by transmission electron microscopy to mold themselves into the fissures and folds of the outer layer of the guard cells (Fig. 7) (Mendgen, 1973a), it is still possible that the fungus is responding to some much smaller feature of the stomatal surface.

The molecular events involved in this perception of surface stimuli are unknown, both for directional growth and appressorium induction. Assuming that Dickinson's (1977) claim is correct that surface contact changes the orientation of the microfibrils in the fungal wall, one still has to explain how such changes induce the physiological events which must take place within the cytoplasm. Thus it is intriguing that in the one instance where a young appressorium of *Uromyces phaseoli* var. *vignae* was accidentally sectioned at right angles to the membrane on which it was growing, the fungal wall is much thinner where it touches the membrane and the plasmalemma in this region appears more electron opaque (Fig. 8) (Heath and Heath, unpublished). The generality of this observation needs to be established,

---

Fig. 6.   Scanning electron micrograph of an appressorium (A) of *Uromyces phaseoli* var. *typica* formed over the edge of a crater formed by spraying a collodion membrane with ethyl acetate. (× 3,300) (From Wynn, 1976.)

Fig. 7.   Thin section through an appressorium (A) of *U. phaseoli* var. *typica* over a stoma of *Phaseolus vulgaris*. Note the way in which the lower surface of the appressorium has molded itself into a fold on the guard cell surface (arrow). (× 2,400) (From Mendgen, 1973a.)

Fig. 8.   Thin section through an appressorium (A) of *U. phaseoli* var. *vignae* formed on an oil-containing collodion membrane (single arrow). Note that the fungal wall (double arrows) is much thinner where it contacts the membrane. (× 27,400) (M. C. Heath and I. B. Heath, unpublished.)

however, but it suggests that more detailed ultrastructural work on the responses of rusts to surface stimuli could be worthwhile.

III. DEVELOPMENT OF INFECTION STRUCTURES

Many fungal pathogens have to undergo morphological changes before they can infect a host plant, but commonly these "infection structures" amount to no more than an appressorium (for example, the intracellular structures formed by the basidiospore appressorium more closely resemble later parasitic phases than morphologically distinct structures specialized for infection; thus they will be considered in Chapter 4, Section II). However, the urediospore germ tube goes through a series of complicated morphological events, forming sequentially an appressorium, infection peg, substomatal vesicle, infection hypha, and, by delineation of the tip of the latter by a septum, the first haustorial mother cell (see Fig. 1). Presumably these structures evolved in relation to the stomatal mode of entry into the tissue, but their formation now appears to be mandatory for infection to occur, even if stomatal ingress is rendered unnecessary by removal of the epidermis (Maheshwari, 1966; Rossetti and Morel, 1958). Similarly there is some evidence that infection structures have to form before saprophytic growth of some axenically culturable rusts can take place (Williams, 1971).

Whatever the method used to induce appressoria (see Section II, this chapter) the development of the remaining infection structures seems to follow without the need of additional stimuli. This means that complete infection structures can easily be induced away from the living plant by the use of artificial membranes or, in the case of *Puccinia graminis* var. *tritici*, temperature shock or chemical treatments. Scanning electron micrographs of such artificially induced infection structures (Wynn, 1976; Paliwal and Kim, 1974) show that each rust species has a characteristic, different, external morphology (Figs. 9, 10, 11) which is apparently identical to that seen during host infection. Several studies have exploited this "artificial" induction of infection structures in order to study the accompanying changes in ultrastructure, and such studies of nucleus and nucleolus ultrastructure (Dunkle *et al.*, 1970; M. C. Heath and I. B. Heath, 1978; Manocha and Wisdom, 1971; Robb, personal communication), as well as the ultrastructural aspects of mitosis (Heath and Heath, 1976), are discussed in Chapter 5. The following sections will therefore examine other aspects of infection structure formation and will also consider germ tube development since the ways in which infection structures differ from the germ tube provide a useful indication of the structural and physiological changes necessary for infection to occur.

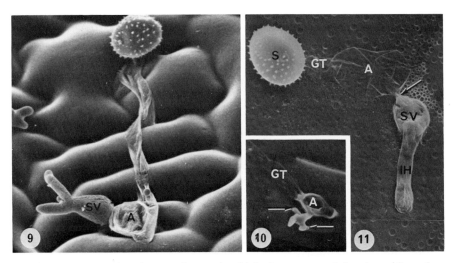

Figs. 9–11. Scanning electron micrographs of infection structures induced away from the living plant.

Fig. 9. Infection structures of *U. phaseoli* var. *typica* formed on a polystyrene replica of a host leaf surface. The appressorium (A) has formed over the guard cells but, since the replica has no stomatal opening, the substomatal vesicle (SV) and characteristically branched infection hypha have formed on the replica surface. (× 671) (From Wynn, 1976.)

Fig. 10. Infection structures of *Puccinia helianthi* induced by an oil-containing collodion membrane (see legend for Fig. 11). Note the characteristic production of four infection hyphae (arrows) from the substomatal vesicle. A, appressorium, GT, germ tube. (× 459) (Freeze-dried, unfixed material, M. C. Heath, unpublished.)

Fig. 11. Urediospore (S), germ tube (GT), appressorium (A), infection peg (arrow), substomatal vesicle (SV), and infection hypha (IH) of *Uromyces phaseoli* var. *vignae* formed on an oil-containing collodion membrane. The craters in the membrane represent the small oil droplets present before the membrane was prepared for electron microscopy; it is thought that undulations caused by these droplets trigger infection structure formation. (× 935) (Freeze-dried, unfixed material, M. C. Heath, unpublished.)

Since most of the information on this topic is only available for *Uromyces phaseoli* var. *vignae* (Heath and Heath, unpublished), it must be borne in mind that other rusts may not necessarily show the same features.

A. Cell Walls and Septa

In *Uromyces phaseoli* var. *vignae* grown on artificial membranes, the 60-nm-thick germ-tube wall is continuous with an inner layer of the urediospore wall which can only be seen around the germ pores (see Fig. 2, and also Fig. 12). Thus the germ-tube wall is not derived from any major part of the spore wall, an observation which correlates with the differences in composition reported for spore and germ-tube walls of the closely related *U. phaseoli* var. *typica* (Trocha and Daly, 1974; Trocha et al., 1974). However, Manocha and Shaw (1967) claim that the germ-tube wall of *Melampsora lini* is continuous with the *outer* layer of the urediospore wall which, if correct, would be a rather unusual situation among the fungi (see review by Akai *et al.*, 1976); however, the poor differentiation of the wall layers in their micrographs makes this conclusion open to question. In *U. phaseoli* var. *vignae*, the synthesis of the germ-tube wall seems to involve fusion of membrane-bound vesicles with the plasmalemma at the hyphal apex, and is discussed in more detail in the following section.

Partial septa, involving an "infolding" of the longitudinal wall (see Chapter 4, Section I,A, and Fig. 11) can be found in germ tubes of both *Uromyces phaseoli* var. *vignae* (M. C. Heath, unpublished) and *Melampsora lini* (Manocha and Shaw, 1967). In the former fungus, these septa occur with no obvious regularity or reason. However, the swelling of the germ-tube tip to form the appressorium is invariably followed by the formation of a septum which delimits the appressorium from the germ tube (see Fig. 2). This septum develops centripetally as the last remnants of the cytoplasm enter the appressorium (M. C. Heath and I. B. Heath, 1978) and serial sections have never revealed any pores once the septum is mature. Like perforate septa in parasitic hyphae (see Chapter 4, Section I,A and Fig. 2), the mature septal wall is composed of two moderately electron-opaque layers. The layer on the appressorium side is continuous with the inner layer of the now *bilayered* appressorial wall while the relationship of the other to the germ-tube wall is less clear (Fig. 13). The fact that the inner appressorial wall layer seems to be an integral part of the septum suggests that it is laid down at the time that the septum forms and such a hypothesis is supported by the absence of this layer during the early stages of appressorial swelling. The addition of this layer results in the appressorium wall being generally thicker (80–100 nm) than that of the germ tube (except where it touches the artificial membrane, see Section II, this chapter).

Figs. 12–16. Morphological changes in wall structure accompanying the formation of infection structures of *Uromyces phaseoli* var. *vignae* on artificial membranes (see Fig. 2, this chapter).

Fig. 12. Germinating urediospore (S) showing that the wall of the germ tube (arrows) can be traced for a short distance along the inside of the spore wall (W), but then seems to disappear. Note the electron-lucent, "foamy" material (P) thought to be storage polysaccharide. (× 17,400) (M. C. Heath and I. B. Heath, unpublished.)

Fig. 13. Junction between an appressorium (A) and a germ tube (G). The appressorial wall seems to be composed of an outer layer which is continuous with the germ-tube wall (large arrow), and an inner layer which is continuous with the wall of the septum (small arrow). Note how the vacuolate appressorium is lined by two membranes, an inner tonoplast and an outer plasmalemma. (× 45,700) (M. C. Heath and I. B. Heath, unpublished.)

Fig. 14. Junction between an appressorium (A) and an infection peg (I). The outer wall layer of the appressorium seems to terminate (arrow) a short distance along the peg and only the inner layer is continuous with the peg wall. (× 30,500) (M. C. Heath and I. B. Heath, unpublished.)

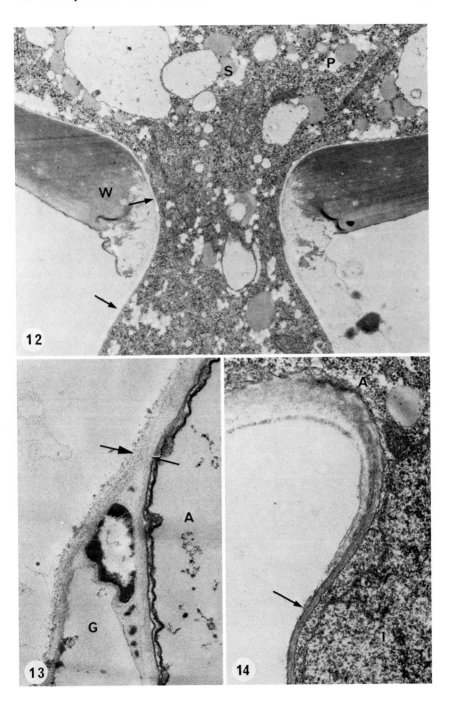

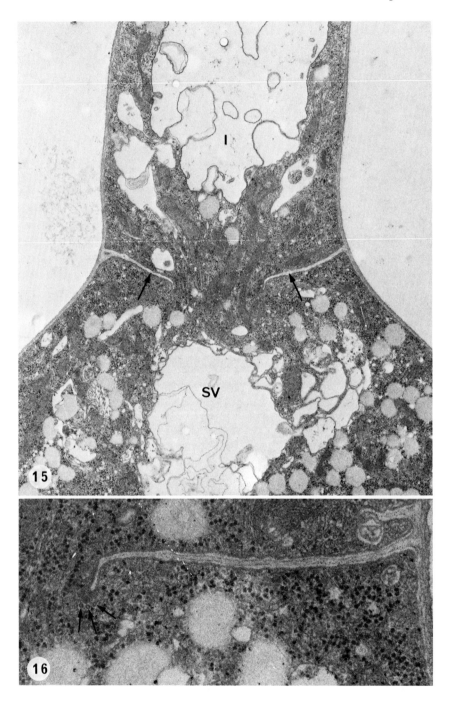

The development of the infection peg of *Uromyces phaseoli* var. *vignae* results from the localized extension of the *inner* layer of the appressorium wall (Fig. 14); thus the infection-peg wall is not continuous with that of the germ tube. However it is continuous with the 40- to 60-nm-thick wall surrounding the substomatal vesicle and infection hypha (Fig. 15) indicating that ultrastructural signs of a change in wall structure do not universally accompany changes in morphology. Significantly, these observations suggest that a change in wall type occurs only as the fungus enters the leaf. This correlates with the observation by Dickinson (1955) that intercellular hyphae of *Puccinia triticina* are more permeable than germ tubes to certain dyes, a difference he attributes to differences in the cell walls. [Such a change in wall type may also explain the quenching of fluorescence observed after dye-labeled basidiospore germ tubes of *Cronartium ribicola* pass between the guard cells of the host (Patton and Johnson, 1970).] In a recent ultrastructural study, Dickinson (1977) also claims that the wall microfibrils of intracellular hyphae of *P. coronata* have a lesser degree of orientation, and are more easily exposed by acid treatment, than those of germ tubes.

All these results suggest that the organization and composition of the germ-tube wall differ from that of infection structures within the plant. This might explain the results of Mayama *et al.* (1975) who could find no detectable increases in glucosamine content following the entry of *P. graminis* f. sp. *tritici* into the leaf. However in *U. phaseoli* var. *vignae*, histological tests for chitin at the light microscope level suggest that this polysaccharide is present in all parts of the infection structures (M. C. Heath, unpublished). It is hoped that the ability to induce more or less synchronous development of infection structures away from the plant will be exploited in the future to determine the exact nature of any temporal changes in wall composition. It would also be of interest to know whether such changes reflect the change in mode of nutrition once the fungus enters the tissue or whether they are merely a response to a lesser need for water conservation. Alternatively (or in addition) such changes may reflect the need of the germ tube, but not of the infection structures, to be able to respond to thigmotropic stimuli on the plant surface (Dickinson, 1977).

From the above-mentioned descriptions, one can see a marked similarity, at least in *U. phaseoli* var. *vignae*, in the way in which the germ tube and in-

Fig. 15. Junction between an infection peg (I) and a developing substomatal vesicle (SV). Note the apparent continuation of the walls of both structures, and the developing septum (arrows). Light microscopy of living cells shows that the completion of the nonperforate septum coincides with the passage of the last remnants of the cytoplasm of the infection peg into the substomatal vesicle. (× 15,800) (I. B. Heath and M. C. Heath, unpublished.)

Fig. 16. A higher magnification of part of the developing septum shown in Fig. 15. Note the electron-opaque region, and the three microtubules (arrows), lining the margin of the septum (× 57,800) (I. B. Heath and M. C. Heath, unpublished.)

fection peg are surrounded by walls continuous with only one of the wall
layers of the urediospore and the appressorium respectively (see Fig. 2).
Similar differential extension of wall layers also occurs during the formation
of the penetration peg of the dikaryotic haustorium (see Fig. 2 and Chapter
4, Section II,A,2) and is characteristic of the germination of other types of
fungal spores as well as the blastosporic mode of spore development (see
review by Akai *et al.*, 1976); thus it appears to be a common method in
fungi of achieving a change in wall structure (and probably composition)
from one morphological form to the next.

The swelling of the tip of the infection hypha of *Uromyces phaseoli* var.
vignae to form a substomatal vesicle has been shown by time-lapse
photography (M. C. Heath, unpublished) to occur extremely rapidly. The
fact that the wall of the developing vesicle is usually much thinner than that
of the infection peg suggests that, as in the formation of the initially spherical
body of the dikaryotic haustorium (see Chapter 4, Section II,A,2), expansion
of the vesicle is accomplished by extension, rather than synthesis, of the cell
wall. As the cytoplasm migrates into the developing vesicle, a septum begins
to develop which will eventually delimit this structure from the infection
hypha (Fig. 2). From the one example of an immature septum examined
ultrastructurally (Fig. 15), it appears that this septum initially develops in the
same manner as developing perforate septa in established mycelium (see
Chapter 4, Section I,A, and Figs. 4 and 5); however, at maturity, serial sections reveal no central pore. During development, a band of electron-opaque
material can be seen in the cytoplasm lining the inner margin of the septum,
and associated with this is a concentric band of microtubules (Fig. 16); the
role of these structures, and whether they are a general feature of the formation of this type of septum, awaits investigation (see also septum formation in
intercellular hyphae, Chapter 4, Section I,A).

As mentioned above, the infection hypha develops from the substomatal
vesicle by localized growth which involves no obvious change in wall structure. Infection structure formation ends, by definition, with the formation of
a septum which delimits the tip of the infection hypha to form a haustorial
mother cell. Details of haustorial mother cell ultrastructure and development
are described in Chapter 4, Section II.

B. Cytoplasmic Changes

The structural organization of the apex of the urediospore germ tube of
Uromyces phaseoli var. *vignae* (Fig. 17) resembles the tip of the basidiospore
germ tube of *Gymnosporangium juniperi-virginianae* (Mims, 1977a) and the
hyphal apices of other septate fungi (see Beckett *et al.*, 1974). Ribosomes and
membrane-bound cisternae occur at the extreme apex but mitochondria,
microbodies, and lipid granules are excluded. Membrane-bound "apical

vesicles" (Grove and Bracker, 1970), approximately 80–150 nm in diameter, are abundant in this region. The tripartite structure of their bounding membrane, the similar electron opacity of their contents to that of the wall, and the presence of electron-opaque material, similar to that seen in some vesicles, within indentations in the plasmalemma, all support the hypothesis that these vesicles contribute membrane and other materials to the growing apex (Grove *et al.*, 1970; Heath *et al.*, 1971). As shown for axenically grown *Melampsora lini* (see Chapter 4, Fig. 13) (Coffey, 1975), the apex contains no specialized region corresponding to the Spitzenkörper (Grove and Bracker, 1970) of some other septate fungi.

In axenically grown *Melampsora lini* (Coffey, 1975) (Chapter 4, Fig. 13) the apical vesicles appear to develop just behind the apex from tubular cisternae resembling those thought to give rise to apical vesicles in other septate fungi (Beckett *et al.*, 1974). Some micrographs (e.g., Fig. 17) suggest a similar origin of vesicles in germ tubes of *U. phaseoli* var. *vignae* but usually vesicle-associated cisternae are seen only in the vicinity of the two nuclei which lie about midway between the apex and the vacuole (M. C. Heath, unpublished). The large distance separating these structures from the apex makes it difficult to determine whether the developing vesicles are indeed the same as those seen at the hyphal tip, although the presence of similar electron-opaque material in both types suggests that this is so.

As might be expected from the almost universal occurrence of apical vesicles during localized wall synthesis in the fungi (Beckett *et al.*, 1974), numerous apical vesicles are also found at the site of urediospore germination in *Puccinia graminis* f. sp. *tritici* (Hess *et al.*, 1975) and *Uromyces phaseoli* var. *vignae* (Fig. 18) (Heath and Heath, unpublished). In this latter fungus, they are less abundant and have a denser matrix at the tips of developing infection pegs or infection hyphae (Fig. 19), which may reflect the much slower growth rate of these structures compared with that of the germ tube (demonstrated by time-lapse photography, M. C. Heath, unpublished). The much more rapid expansion of the substomatal vesicle is also accompanied by no marked accumulation of vesicles near the fungal wall but, as discussed in Section III,A, of this chapter, vesicle expansion is possibly achieved by stretching, rather than synthesis, of the fungal wall.

At all stages of the development of infection structures of *Uromyces phaseoli* var. *vignae*, the general cytoplasm contains the same basic components found in other stages of the life cycle (see Chapter 4, Section I,B) such as mitochondria, microbodies, lipid droplets, microtubules, endoplasmic reticulum, and ribosomes. Clearly recognizable glycogen granules, however, have not been seen, either in infection structures formed on artificial membranes or in those developed in the host; they have, however, been shown in the substomatal vesicle formed within the host by *U. phaseoli* var. *typica* (Mendgen, 1973a). Material with an electron-lucent "foamy" ap-

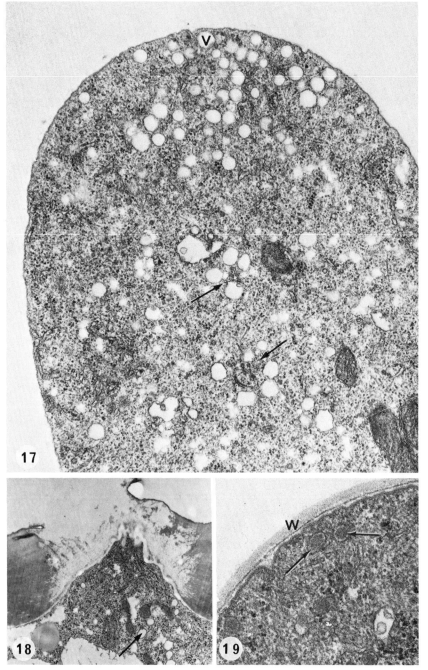

Figs. 17–19. Apical growth of infection structures of *Uromyces phaseoli* var. *vignae*.

pearance is found in the germinating, but not in dormant, urediospores of *Puccinia graminis* f. sp. *tritici* (Sussman *et al.*, 1969) and *U. phaseoli* var. *vignae* (Fig. 20) (Heath and Heath, unpublished). In the latter fungus, this material also occurs in pseudoparenchymatous hyphae of the uredia where it can be removed from the section by α-amylase treatment (M. C. Heath, unpublished). Thus it seems to contain polysaccharide, a conclusion reached for morphologically similar material seen in other fungi (e.g., Cole and Aldrich, 1971, and also references within the review by Smith *et al.*, 1976). In *U. phaseoli* var. *vignae*, the appearance of this material during spore germination and its subsequent reduction in abundance during germ tube growth (Heath and Heath, unpublished) correlates remarkably well with biochemical evidence from *U. phaseoli* var. *typica* that glucomannan proteins are synthesized during the first two hours of germ tube elongation and are then rapidly broken down (Wynn and Gajdusek, 1968). Even the small increase in mannan synthesis which was observed at the time of substomatal vesicle formation (Wynn and Gajdusek, 1968) seems to correlate with the observation that the developing infection peg and substomatal vesicle of *U. phaseoli* var. *vignae* often contain slightly more of this electron-lucent material than the appressorium or the infection hypha.

Biochemical evidence from several rusts suggests that lipids are also utilized during urediospore germination (see review by Hess and Weber, 1976). Correspondingly, Williams and Ledingham (1964) have claimed that the "oil bodies" of *Puccinia graminis* f. sp. *tritici,* as seen by transmission electron microscopy, are smaller in the germ tube than in the urediospore. However, no obvious change in numbers or size of liquid droplets can be observed in *Uromyces phaseoli* var. *vignae,* either during germination of urediospores or during the development of infection structures, except for a slight increase in numbers in the substomatal vesicle and infection hypha when these form away from the host plant (M. C. Heath, unpublished). The abundance of lipid in the substomatal vesicle has also been noticed for *Puccinia graminis* f. sp. *tritici* (Skipp *et al.*, 1974), *P. helianthi* (M. C. Heath, unpublished), and

---

Fig. 17.   Apex of a germ tube showing apical vesicles (V). Membrane configurations (arrows) suggestive of the origin of these vesicles from single tubular cisternae can be seen just behind the apex (compare with Chapter 4, Fig. 13), but usually such profiles are seen only in the region of the nuclei. The fungal wall of the germ tube stains poorly and is therefore almost invisible. ($\times$ 27,800) (I. B. Heath and M. C. Heath, unpublished.)

Fig. 18.   Germinating urediospore. Vesicles (arrow) resembling those shown in Fig. 17 can be seen in the cytoplasm in the vicinity of the germ pore. ($\times$ 15,400) (I. B. Heath and M. C. Heath, unpublished.)

Fig. 19.   Apex of an infection hypha. Apical vesicles (arrows) are less abundant than in the germ tube (see Fig. 17) and also have more densely staining matrix. Note that the fungal wall (W) also stains more strongly than that of the germ tube. ($\times$ 57,800) (I. B. Heath and M. C. Heath, unpublished.)

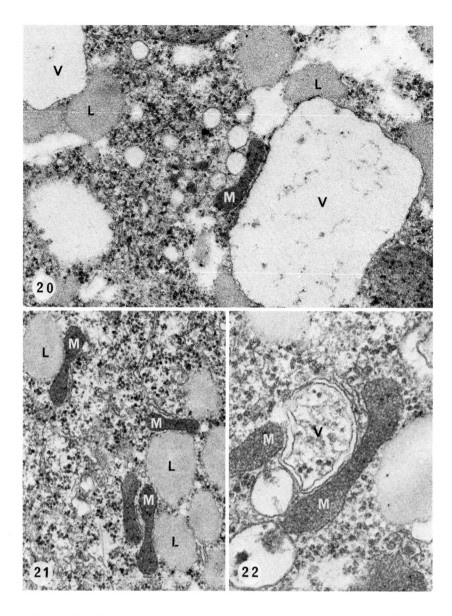

Figs. 20–22. Changes in the association of microbodies with various structures during the formation of infection structure of *Uromyces phaseoli* var. *vignae* on artificial membranes.

Fig. 20. Association of vacuoles (V) with lipid droplets (L) and microbodies (M) within a germinating urediospore. (× 45,400) (M. C. Heath and I. B. Heath, unpublished.)

Fig. 21. Association of microbodies (M) with lipid droplets (L) in a substomatal vesicle. (× 37,400) (M. C. Heath and I. B. Heath, unpublished.)

Fig. 22. Association of microbodies (M) with what seems to be an autophagic vacuole (V) in an infection hypha bearing a haustorial mother cell and secondary hypha. (× 74,700) (M. C. Heath and I. B. Heath, unpublished.)

U. phaseoli var. *typica* (Mendgen, 1973a), all growing in the host. In both *U. phaseoli* var. *typica* (Mendgen, 1973b) and *U. phaseoli* var. *vignae* (Heath and Heath, unpublished), these lipid droplets are often associated with microbodies (Fig. 21) and Mendgen (1973b) has suggested that the latter may be involved in lipid breakdown and could therefore be classified as glyoxysomes. However, he found the diaminobenzidine (DAB) test for the activity of catalase (the marker enzyme for glyoxysomes) to be negative, although evidence of catalase activity was found when hydrogen peroxide was added to germ tube homogenates. The latter observation supports the suggestion by Maxwell *et al.* (1977) that this negative DAB test, and that reported for microbodies in intercellular mycelium (see Chapter 4, Section I), is most likely due to inappropriate test conditions, rather than an absence of catalase. Further support for this suggestion is the biochemical demonstration that urediospores of several rust species contain key glyoxylate pathway enzymes (see review by Reisener, 1976). Interestingly, these substomatal vesicle microbodies of *U. phaseoli* var. *vignae* (M. C. Heath, unpublished) and *U. phaseoli* var. *typica* (Mendgen, 1973a) only rarely contain the crystal lattice, commonly thought to be rich in catalase, which is seen in most microbodies associated with perforate septa (see Chapter 4, Section I,A); conceivably this indicates that the enzyme in the substomatal vesicle microbodies is in a more soluble, and more readily available, form. It is also of interest that in *U. phaseoli* var. *vignae*, microbodies do not seem to be associated with lipid droplets in germinating urediospores (where microbodies and lipid droplets are *independently* associated with vacuoles, Fig. 20), germ tubes or appressoria (Heath and Heath, unpublished). By analogy with other systems where changes in microbody associations have been correlated with biochemical activity (Trelease *et al.*, 1971), this lack of association may indicate a corresponding lack of extensive lipid utilization during these phases of growth. Similarly, a further change in microbody function is indicated by a new type of association formed during the formation of haustorial mother cells and secondary hyphae (i.e., the first hyphal branch to develop from the infection hypha) on artificial membranes; the microbodies now become appressed to what appear to be autophagic vacuoles (Fig. 22), and may therefore have a lysosomal function. The lack of lipid droplets at this stage, which also occurs in *U. phaseoli* var *typica* in the host (Mendgen, 1973a), suggests that such autolysis may be related to fungal senescence induced by the exhaustion of endogenous nutrient sources and the absence of any exogenous materials from a host plant.

C. Cytoplasmic Migration

On artificial membranes, urediospore germination and infection structure development by *Uromyces phaseoli* var. *vignae* are accompanied by no obvious change in the total volume of the cytoplasm. This suggests that, like the

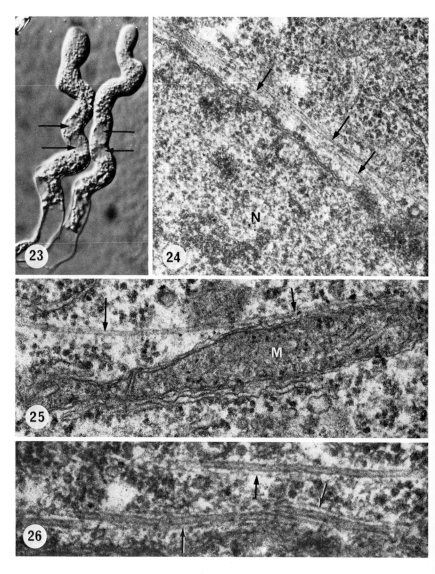

Figs. 23–26. Cytoplasmic migration during the formation of infection structures of *Uromyces phaseoli* var. *vignae* on artificial membranes.

Fig. 23. Light micrograph of two germ tubes showing the consistent positioning of the nuclei (arrows) about midway between the apex and the vacuole. (× 666) (M. C. Heath, unpublished.)

Fig. 24. Association of microtubules (arrows) with a nucleus (N) during migration of the cytoplasm from an appressorium to a substomatal vesicle. As in the germ tube, the nuclei always migrate with the middle third of the cytoplasm. (× 60,200) (From I. B. Heath and M. C. Heath, 1978.)

closely related *U. phaseoli* var. *typica* (see review by Lovett, 1976), little net protein synthesis occurs during these processes. Thus the cytoplasm must migrate from one structure to the next as each one is formed (I. B. Heath and M. C. Heath, 1978) and the few available ultrastructural studies of early stages of infection suggest such migration is also typical of infection structures in the host plant (Locci and Bisiach, 1970; Mendgen, 1973a). Direct observation and time-lapse photography of living infection structures of *U. phaseoli* var. *vignae* grown on artificial membranes, show three categories of cytoplasmic movement during this migration (a) a general migration of cytoplasm and organelles during which the nuclei, in particular, maintain their characteristic position with remarkable constancy (Fig. 23), (b) a relatively slow displacement of various organelles and unidentified particles relative to one another and the growing region of the fungus, and (c) the rapid, erratic, short distance saltations of particles and organelles with the exception of elongate mitochondria. Electron microscopy reveals that during migration, nuclei and elongate mitochondria are commonly associated with laterally placed cytoplasmic microtubules (Figs. 24 and 25) whereas lipid droplets, identified as being the major component of the population of saltating particles seen with the light microscope, are not (I. B. Heath and M. C. Heath, 1978). A tempting conclusion from these observations is that microtubules are involved in the maintenance of position of these two organelles within the migrating cytoplasm and may also control their seemingly nonrandom displacements. In support of this hypothesis, antimicrotubule agents have a greater disruptive effect on the position and movements of nuclei and mitochondria than on saltations of lipid droplets (F. Herr and M. C. Heath, unpublished).

Serial section analysis shows that the majority of these cytoplasmic microtubules are less than 2 μm long which, assuming that this is not an artifact of fixation, seems to preclude their acting as a rigid cytoskeleton to which the organelles can become anchored (I. B. Heath and M. C. Heath, 1978). Thus Heath and Heath suggest that the microtubules may act as intermediaries between the organelles and some other extensive cytoskeletal system which is not preserved, or is rendered invisible, by standard fixation procedures. Such a cytoplasmic matrix has been suggested to exist in other organisms (see I. B. Heath and M. C. Heath, 1978; Pollard, 1976; and references contained therein) and possible evidence for its presence in *Uromyces phaseoli* var. *vignae* is the observation of 3- to 6-nm-diameter, ac-

---

Fig. 25. Association of a microtubule (arrows) with a mitochondrion (M) during cytoplasmic migration from an appressorium to a substomatal vesicle. ($\times$ 72,800) (I. B. Heath and M. C. Heath, unpublished.)

Fig. 26. Filaments (arrows) sometimes seen associated with microtubules in infection structures. ($\times$ 115,400) (From I. B. Heath and M. C. Heath, unpublished.)

tinlike filaments associated with some of the cytoplasmic microtubules (Fig. 26). Such filaments may represent components of a much more extensive system stabilized by their association with the microtubules against the known disruptive effects of fixation.

D. Vacuolation

As the cytoplasm moves from one portion of the developing infection structures to the next, a vacuole develops in the vacated portion of the fungus. (This is a practical necessity since, if the whole protoplast moved within the cell wall, there would be no mechanism to protect the naked plasmalemma at the "rear end" of the fungus against osmotically induced swelling of the protoplast.) Thus the "empty" portions of the infection structures are in fact lined by an extremely thin layer of cytoplasm which commonly consists only of the plasmalemma and the tonoplast (Fig. 13).

The presence of both plasmalemma and tonoplast in vacated portions of the infection structures means that a large portion of the metabolic activity of the fungus must be directed toward rapid membrane synthesis corresponding to the rate of fungal growth. The additional plasmalemma thus needed can be accounted for, at least in the germ tube, infection peg, and infection hypha, by the addition of membrane from the apical vesicles to the growing region (see Chapter 4, Section I,B). The origin of the tonoplast, however, has been little studied in the rusts. From what little evidence that exists for other fungi there seem to be at least two mechanisms of vacuole formation; in the oomycete *Saprolegnia* spp., the vacuole in a number of stages of the life cycle seems to develop by the fusion of smaller vesicles containing a characteristic electron-opaque body often with a fingerprint-like substructure (Gay *et al.*, 1971), while in the ascomycete *Neurospora crassa,* the membrane of young vacuoles seems to be derived from the endoplasmic reticulum (Bracker, 1974). In infection structure formation in the rusts, the delineation of the appressorium and the substomatal vesicle by a septum results in three stages of vacuole formation which are (a) vacuole formation in the urediospore which

---

Figs. 27–29. Vacuole formation in the appressorium formed by *Uromyces phaseoli* var. *vignae* on artificial membranes.

Fig. 27. A newly formed appressorium showing large numbers of "dense-body" vesicles (arrows) and larger vacuoles (V) which seem to coalesce to form one large vacuole as the cytoplasm migrates into the infection peg and substomatal vesicle. (× 3,800) (I. B. Heath and M. C. Heath, unpublished.)

Fig. 28. A vacuole (V) in a young appressorium with what appears to be a "dense body" (arrow) within it. (× 74,400) (I. B. Heath and M. C. Heath, unpublished.)

Fig. 29. A dense-body vesicle (D) next to a larger vacuole (V). Note that both are bounded by membranes of similar thickness and appearance. (× 158,400) (I. B. Heath and M. C. Heath, unpublished.)

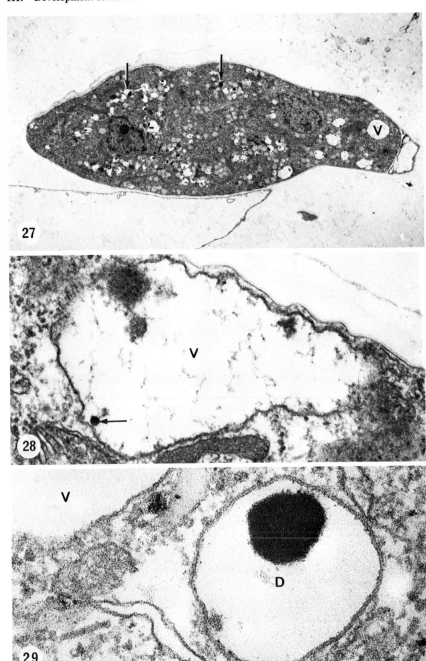

continues into the germ tube as the latter grows in length, (b) vacuole formation in the appressorium as the cytoplasm moves into the developing
substomatal vesicle, and (c) vacuole formation in the substomatal vesicle as
the infection hypha elongates. The second stage has been examined
ultrastructurally in detail for *Uromyces phaseoli* var. *vignae* (Heath and
Heath, unpublished) and in this situation, the developing vacuole contains
electron-opaque bodies (but with *no* fingerprint-like substructure) which appear identical to those observed in smaller vesicles distributed throughout the
cytoplasm (Figs. 27–29). The association of some of these vesicles with the
developing tonoplast, and the close resemblance between the membranes of
the two structures, is consistent with the hypothesis that the vacuole enlarges
by fusion with these vesicles. However, if this hypothesis is correct, then only
a small proportion of the "dense-body" vesicles of the appressorium are involved in vacuole formation (or there is concomitant synthesis of these
vesicles), since many can still be seen in the young substomatal vesicle and the
developing infection hypha. Whether they are involved in subsequent
vacuolation of the substomatal vesicle has not been determined but no
dense-body vesicles can be found by the time the first haustorial mother
cell and secondary hypha develop (Heath and Heath, unpublished).

Ultrastructural evidence, similar to that described above for the appressorium, suggests that the dense-body vesicles are also involved in
tonoplast formation in the germinating urediospore but these vesicles have
never been seen near the vacuolating portions of the germ tube or
intercellular hyphae; nor do they occur during vacuolation of the haustorial
mother cell (see Chapter 4, Section II,A,2). Thus their involvement in
vacuolation does not seem to be mandatory and it is likely that these vesicles
have some additional function for which tonoplast synthesis may be only in-
cidental. One possibility is that these vesicles contain material whose
discharge into the vacuole increases the osmotic pressure and, by the
resulting increase in water uptake from the environment, increases the
hydrostatic pressure. The limited occurrence of this process, being restricted
to the urediospore and appressorium, may relate to the need for particularly
high hydrostatic pressures within these cells to (a) rupture the outer walls of
these structures (see Chapter 2, Section V,D), and/or (b) to force the infection peg through a possibly closed stoma, and/or (c) to produce rapid expansion of the substomatal vesicle. It is interesting that morphologically similar
dense-body vesicles can be seen in micrographs of many different fungi in
various stages in the life cycle (Beckett *et al.*, 1974); whether they all have a
role similar to that suggested here remains to be determined.

4

Vegetative Growth in the Susceptible Host

I. INTERCELLULAR HYPHAE

Only vegetative mycelium, unassociated with haustorium or spore formation, will be covered in this section. The development and structure of the haustorial mother cell, since it is so closely linked to haustorium formation, will be considered in Section II.

A. Cell Walls and Septa

The cell walls of intercellular rust mycelium characteristically appear to be composed of two, moderately electron-opaque, fibrillar layers (Fig. 1), although this appearance does seem to vary with different fixation and staining techniques. Commonly an additional amorphous outer layer (Fig. 1) is also present in both monokaryotic hyphae (*Cronartium ribicola,* Boyer and Isaac, 1964; Robb *et al.,* 1973; Welch and Martin, 1975; *Peridermium pini,* Walles, 1974; *Puccinia sorghi,* Rijkenberg and Truter, 1973a) and dikaryotic hyphae (*Hemileia vastatrix,* Rijkenberg and Truter, 1973a; Rijo and Sargent, 1974; *Uromyces phaseoli* var. *typica,* Müller *et al.,* 1974a) which has been suggested to have a role in the adhesion of the hyphae to other hyphae or host cells (Hardwick *et al.,* 1971; Müller *et al.,* 1974a; Rijkenberg and Truter, 1973a; Walles, 1974; Welch and Martin, 1973). Welch and Martin (1974) also suggest that C. *ribicola* obtains host nutrients at the sites of host-hyphae adhesion but Boyer and Isaac (1964) suggest that this outer layer protects the fungus against toxic host substances. Presumably, this layer is secreted by the fungus, but whether it should be regarded as part of the wall is debatable until its composition is known.

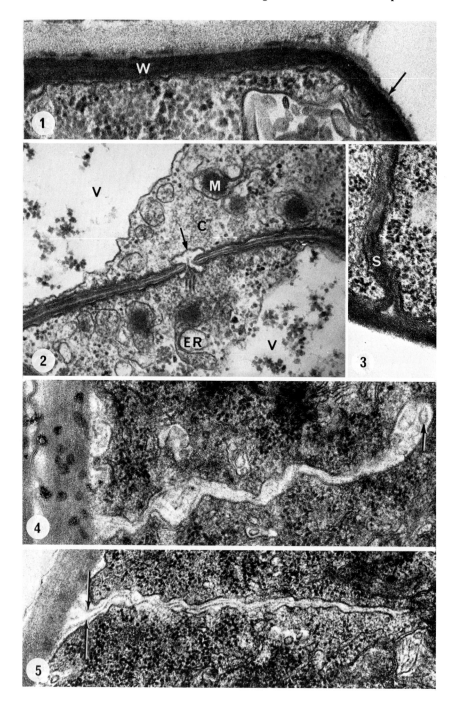

As expected for members of the basidiomycetes, intercellular hyphae of all rusts are septate. However, electron microscopy has revealed three basic types of septa in the Uredinales (Littlefield, 1971a; Littlefield and Bracker, 1971a), and of these, the perforate type (Fig. 2) is the one usually responsible for partitioning the vegetative intercellular mycelium into uninucleate or binucleate portions. This septum, commonly considered to be the "typical" rust septum, develops centripetally (*Melampsora lini,* uredial stage, Littlefield and Bracker, 1971a; *Puccinia sorghi,* aecial stage, Rijkenberg and Truter, 1974b; *Puccinia* spp., uredial stage, Harder, 1976b; *Uromyces phaseoli* var. *typica,* uredial stage, Müller *et al.,* 1974a), and early stages in its formation can be recognized as invaginations of the plasmalemma toward the center of the hypha (Fig. 4). The region within the invagination is at first electron-lucent (Littlefield and Bracker, 1971a; Harder, 1976b) and sometimes contains vesicular profiles (Fig. 4) (Littlefield and Bracker, 1971a) but later two electron-opaque lamellae appear which become progressively thicker and are continuous with the inner layer of the longitudinal wall (Fig. 5). During their development a region of differentiated cytoplasm occurs along the *faces* of the developing septum of *Melampsora lini* (Fig. 5) (Littlefield and Bracker, 1971a) comparable to that seen near the *margins* of developing partial septa of *U. phaseoli* var. *vignae* (see following paragraph). No electron-opaque ring or associated microtubules, such as those seen during development of the nonperforated septum in *U. phaseoli* var. *vignae* in-

---

Fig. 1. Wall (W) of an intercellular hypha of *Uromyces phaseoli* var. *vignae* (uredial stage). Two fibrillar layers and an outer amorphous layer can be distinguished. (× 64,600) (M. C. Heath, unpublished.)

Fig. 2. Typical septal pore apparatus of *U. phaseoli* var. *vignae* (uredial stage) showing the diaphragms (arrow) covering the pore, and the surrounding differentiated region of cytoplasm (C) bounded by crystal containing microbodies (M). Each microbody is characteristically associated with a cisterna of endoplasmic reticulum (ER). The fibrillar material seen on the lower side of the pore is not always present. Note the vacuoles (V) on either side of the septum, which are characteristic of older regions of the mycelium. (× 47,300) (M. C. Heath, unpublished.)

Fig. 3. The edge of a typical perforate septum (S) of *U. phaseoli* var. *vignae* (uredial stage) showing the continuation of the two septal lamellae with the *inner* layer of the longitudinal wall. (× 57,600) (I. B. Heath, unpublished.)

Fig. 4. Part of a developing, perforate septum of *Melampsora lini* (uredial stage). The region within the invaginated plasmalemma is filled with diffuse, slightly-stained material and several vesicles (arrow). (× 70,000) (From Littlefield and Bracker, 1971a. Reproduced by permission of Cambridge Univ. Press from *Trans. Br. Mycol. Soc.* **56,** 181-188.)

Fig. 5. Part of a developing, perforate septum of *Melampsora lini* (uredial stage) at a slightly later stage of development than that in Fig. 4. Two thin electron-opaque lamellae (arrows), continuous with the longitudinal wall, can now be discerned within the invaginated plasmalemma. Note also the region of ribosome-free cytoplasm surrounding most of the septum. (× 50,000) (From Littlefield and Bracker, 1971a. Reproduced by permission of Cambridge Univ. Press from *Trans. Br. Mycol. Soc.* **56,** 181-188.)

fection structures (see Chapter 3, Section III,A, and Fig. 16) were observed in the vicinity of developing septa of *M. lini,* but one reason for this could be that the *M. lini* infections were fixed in the cold, a treatment known to disrupt cytoplasmic microtubules. During and after their formation, the two septal layers are separated by an electron-lucent region which extends from the margin of the central pore to the inner limit of the *outer* layers of the longitudinal wall (Figs. 3, 4, 5).

The pore in the mature septum can vary in diameter in one mycelium (e.g., values of 30 to 150 nm have been recorded for *U. phaseoli* var. *vignae,* M. C. Heath, unpublished) but usually seems to be less than 100 nm and may be as little as 15–30 nm (*M. lini,* Littlefield and Bracker, 1971a). Characteristically the pore is associated with a "septal pore apparatus" which is illustrated in Fig. 2 (for a diagrammatic representation see Fig. 19, Section II,A, this chapter). Covering the pore on both sides of the septum are thin electron-opaque diaphragms which extend over the surface of the plasma-lemma for a short distance (e.g., 150 nm in *M. lini,* Littlefield and Bracker, 1971a; 300 nm in *U. phaseoli* var. *vignae,* Heath and Heath, 1975) on either side of the pore. Characteristically, the cytoplasm next to the diaphragms is differentiated to form an organelle-free hemisphere on either side of the septum. Crystal-containing microbodies (see Section I,B, this chapter) are characteristically found around the periphery of this region, each usually enfolding a cisterna of what appears to be endoplasmic reticulum (Coffey *et al.,* 1972a; Heath and Heath, 1975).

This distinctive septal pore apparatus has been clearly observed in the intercellular dikaryotic mycelium of *Melampsora lini* (Coffey *et al.,* 1972a; Littlefield and Bracker, 1971a), *Puccinia helianthi* (Coffey *et al.,* 1972a), *Puccinia coronata* (Harder, 1976b), *Uromyces phaseoli* var. *typica* (Müller *et al.,* 1974a), and *Uromyces phaseoli* var. *vignae* (M. C. Heath, unpublished) as well as in monokaryotic hyphae of *P. sorghi* (Rijkenberg and Truter, 1974b, 1975), *P. coronata* (Harder, personal communication) and *U. phaseoli* var. *vignae* (D. M. Tighe and M. C. Heath, unpublished). This septal pore apparatus has also been seen in axenically grown dikaryotic hyphae of *P. graminis* f. sp. *tritici* (Harder, 1976b), *U. dianthi* (Jones, 1973) and monokaryotic hyphae of *Cronartium ribicola* grown on pine callus tissue (Robb *et al.,* 1973) and thus seems to be ubiquitous among the rusts, regardless of nuclear condition or type of growth. Nevertheless, several authors have reported seeing structural variations in the basic apparatus, especially in older portions of the mycelium (Coffey *et al.,* 1972a; Jones, 1973; Littlefield and Bracker, 1971a) and in *U. phaseoli* var. *vignae* (M. C. Heath, unpublished), at least half a dozen major or minor deviations from the basic structure have been observed without any obvious correlation with state or age of the surrounding region of the hyphae (Figs. 6 and 7). [This may account for the absence of some characteristic features of the pore apparatus from early micrographs of these structures (e.g., Ehrlich *et al.*, 1968a; Moore, 1963a) but an alternative explanation is the differential preservation of certain features by the permanganate fixation used in these studies.] As found in other septate fungi (Heath, 1975), one of the most commonly reported variations is the occlusion of the pore by electron-opaque material with the same pulley-wheel shape as that outlined by the diaphragms in the more "typical" apparatus (Fig. 8) (Harder, 1976b; Jones, 1973; Littlefield and Bracker, 1971a; Longo and Naldini, 1970; Mims and Glidewell, 1978). Jones (1973) postulates that this plug results from the discharge of the contents of the peripheral microbodies into the pore area but the evidence for this is equivocal and similar plugs develop in haustorial mother cell septa without detectable microbody involvement (see Section I,B, this chapter). Plugging of the pore has also been reported in the hyphae of *Peridermium pini* (Walles, 1974) and *C. ribicola* (Robb *et al.*, 1973) but in these cases the plug sometimes appears more like an inflated dumbbell (Fig. 9), and Walles suggests that it represents two Woronin bodies united inside the pore. The fact that Woronin bodies are generally considered unique to ascomycetes (Bracker, 1967), however, makes this interpretation open to question.

Whether the plugged pore actually restricts intercellular movement, as suggested for plugged pores in other fungi (Bracker, 1967), is also an open question since Jones (1973) has shown a large, greatly constricted, vesicle which appears to have been passing through a "plugged" pore of *Uromyces dianthi* at the time of fixation. Indeed, the physiological role of the "normal" septal pore apparatus in vegetative mycelium is a complete mystery although Ehrlich *et al.* (1968a) have suggested that, through prevention of intercellular migration of nuclei, its function may be to maintain the uninucleate or binucleate condition of each cell. Such a hypothesis has also been postulated for the different, but equally complex, septal pore apparatus of the homobasidiomycetes (Moore, 1965) and is supported by the dissolution of the pore during nuclear migration following dikaryotization in these fungi and by the observation that dissolution of septa, although not necessarily at the pore region, accompanies dikaryotization in the developing aecium of *Puccinia sorghi* (Fig. 10) (Rijkenberg and Truter, 1975). However, one can imagine that nuclear migration could be restricted by a much simpler structure than the one observed and it seems likely that this is not the sole function of the pore apparatus; thus a complete understanding of its role awaits discovery.

Although the septal pore apparatus of the rusts has been suggested to be intermediate between the more simple perforate septum of the ascomycetes and the complex structure found in the homobasidiomycetes (Ehrlich *et al.*,

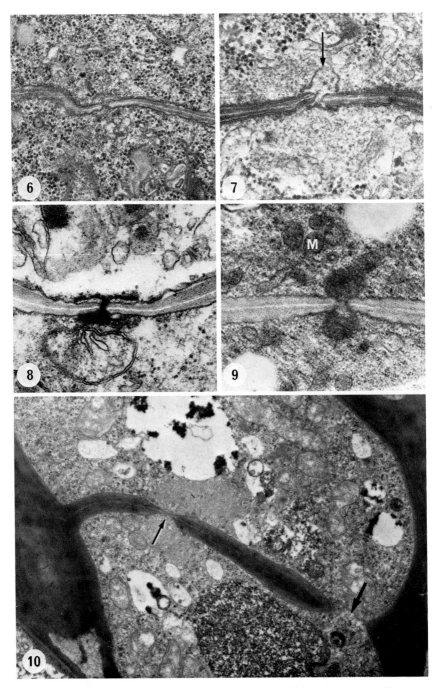

Figs. 6–9. Variation in the septal pore apparatus seen within vegetative mycelium.

1968a), in the authors' opinion, the apparatus of the rusts and the "higher" basidiomycetes are equally complex. Thus any evolutionary derivation of one from the other seems unlikely and it appears more reasonable to suggest that both evolved independently. It is interesting that species of *Septobasidium*, a genus sometimes suggested to be related to the rusts (Couch, 1937), seem to manage with a much more simple, ascomycete-like septum lacking diaphragms, differentiated cytoplasm, or microbodies (Dykstra, 1974). Whether this septum represents a more primitive state, or a secondarily reduced form, is a matter of speculation.

The second type of septum found in the Uredinales, the complete (nonperforated) septum, seems to be more characteristic of pseudoparenchyma of uredial primordia (*Melampsora lini*, Littlefield and Bracker, 1971a; *Puccinia podophylli*, Moore, 1963a) and infection structures (see Chapter 3, Section III,A) than vegetative hyphae, but apart from the absence of the pore and the pore apparatus, this septum resembles the perforate form. However, the third type of septum is both morphologically and developmentally distinct from the other two in that it appears as an extension of *all* the layers of the longitudinal wall and typically has a large, often acentric, pore with no plug or pore apparatus (Fig. 11). These "pseudosepta" (Ehrlich *et al.*, 1968a), "partial septa" (Littlefield and Bracker, 1971a), or "infolded walls" (Rijkenberg and Truter, 1975) are generally found in regions where large numbers of hyphae are crowded into small areas such as in primordial regions of aecia, uredia, and telia (Littlefield and Bracker, 1971a; Mims and Glidewell, 1978; Moore, 1963a; Müller *et al.*, 1974a; Rijkenberg and Truter, 1974b) but are also found elsewhere in infected tissue (Ehrlich *et al.*, 1968a; Müller *et al.*, 1974a; M. C. Heath, unpublished) as well as in germ tubes of *Melampsora lini* (Manocha and Shaw, 1967) and *Uromyces phaseoli* var. *vignae* (M. C. Heath, unpublished). Light microscopy of living germ tubes of

---

Fig. 6. Perforate septum of *Uromyces phaseoli* var. *vignae* (uredial stage); note lack of typical septal pore apparatus. ($\times$ 36,000) (M. C. Heath, unpublished.)

Fig. 7. Evagination of one of the diaphragms (arrow) in an otherwise normal septal pore apparatus of *U. phaseoli* var. *vignae* (uredial stage). ($\times$ 50,000) (M. C. Heath, unpublished.)

Fig. 8. Occluded pore in an intercellular hypha of *Melampsora lini* (uredial stage). The surrounding septal pore apparatus seems disorganized. ($\times$ 26,000) (Littlefield and Bracker, 1971a. Reproduced by permission of Cambridge Univ. Press from *Trans. Br. Mycol. Soc.* **56**, 181–188.)

Fig. 9. Plugged pore of *Cronartium ribicola* (pycnial/aecial stage) growing on *Pinus monticola* tissue cultures where the microbodies (M) have invaded the region of differentiated cytoplasm. Note the different shape of the plugs in Figs. 8 and 9. ($\times$ 54,700) (From Robb *et al.*, 1973. Reproduced by permission of the National Research Council of Canada from *Can. J. Bot.* **51**, 2301–2305.)

Fig. 10. Septum of a fusion cell in the aecium of *Puccinia sorghi* showing the dissolution of the septum (large arrow) away from the region of the septal pore apparatus (small arrow). ($\times$ 12,900) (From Rijkenberg and Truter, 1975.)

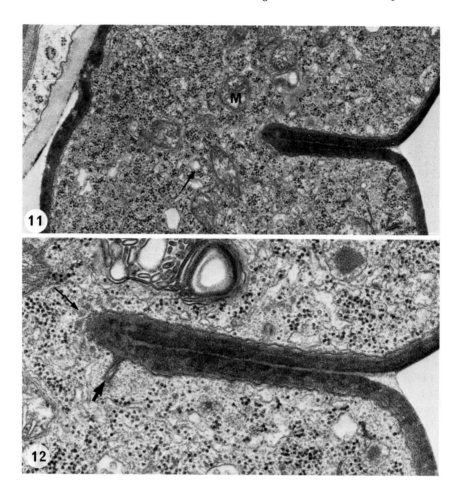

Figs. 11–12.   Partial septa of *Uromyces phaseoli* var. *vignae* (uredial stage).

Fig. 11.   A (possibly) developing partial septum. Note the large size of the pore (which will often remain about this size), its asymmetry within the septum, and the way in which the septum is continuous with all layers of the longitudinal wall. The surrounding cytoplasm contains mitochondria (M) and abundant, rather vesicular, endoplasmic reticulum (arrow). (× 23,000) (M. C. Heath, unpublished.)

Fig. 12.   Detail of a serial section to that shown in Fig. 11. Note the cluster of electron-opaque granules (small arrow) (shown to be spherical by serial sections) near the inner margin of the septum, as well as invaginations of the plasmalemma (large arrow) in this region. Both types of structures are associated with ribosome- and other organelle-free regions of the cytoplasm; such regions do not occur near the outer half of the septum. The tubular, membranous complex above the septum is one of many types of configurations found in rust fungi (see also Fig. 15).(× 47,800) (M. C. Heath, unpublished.)

the latter rust suggests that these partial septa develop centripetally as do the other types and that they do not impede the flow of cytoplasm and organelles when mature (see Chapter 3, Section III,A). Ultrastructural studies of intercellular vegetative hyphae (uredial stage) of this same rust have shown 30–50 nm diameter granules, associated with a differentiated region of the cytoplasm, around the inner margins of what appear to be developing partial septa (Fig. 12). In size and morphology, these vesicles resemble the chitin synthetase particles (chitosomes) seen in *Mucor rouxii* (Bracker *et al.*, 1976) but their exact nature awaits investigation. It is also of interest that if the interpretation of these micrographs is correct, partial septa develop by progressive wall synthesis at the inner edge of the septum accompanied by the simultaneous invagination of the plasmalemma. This is in marked contrast to the perforate septum which develops first as an invagination of the plasmalemma and wall synthesis occurs subsequently over the whole septum surface.

B.   Cell Contents

The ultrastructure of intercellular hyphae has not been studied extensively but all the evidence suggests that the protoplasm generally resembles that of the hyphae of other septate, parasitic or nonparasitic fungi (for basic reviews of hyphal ultrastructure see Bracker, 1967, and Beckett *et al.*, 1974).

The ultrastructure of the hyphal apex within the host has been illustrated only for *Melampsora lini* (Coffey, 1975) and for this species, it seems similar to that of axenically cultured hyphae (Fig. 13) (Coffey, 1975), or germ tubes of *Uromyces phaseoli* var. *vignae* (see Chapter 3, Fig. 17) and *Gymnosporangium juniperi-virginianae* (Mims, 1977a) except that fewer membrane bound "apical vesicles" were observed. In other fungi, these vesicles are assumed to contribute membrane and other materials to the expanding tip surface by fusion with the plasmalemma (Grove *et al.*, 1970; Heath *et al.*, 1971) and their low abundance in parasitic *Melampsora lini* hyphae may indicate a slow rate of growth (Coffey, 1975); however this may be true only of the particular mycelium examined. In parasitic, and axenically grown (Fig. 13) hyphae of *M. lini*, as in other fungal hyphae, large organelles are absent from the extreme apex, but ribosomes, and tubular and vesicular endoplasmic reticulum may all be present as they are in other portions of the mycelium. As expected from light microscope studies (e.g., Savile, 1939), two nuclei per cell are commonly found in dikaryotic hyphae away from the immediate apex (Coffey *et al.*, 1972; Manocha and Shaw, 1967; Müller *et al.*, 1974a; Rijo and Sargent, 1974) but only one in monokaryotic mycelium (Walles, 1974); details of nuclear structure are discussed in Chapter 5. Mitochondria typically have the platelike cristae characteristic of non-oomycetous

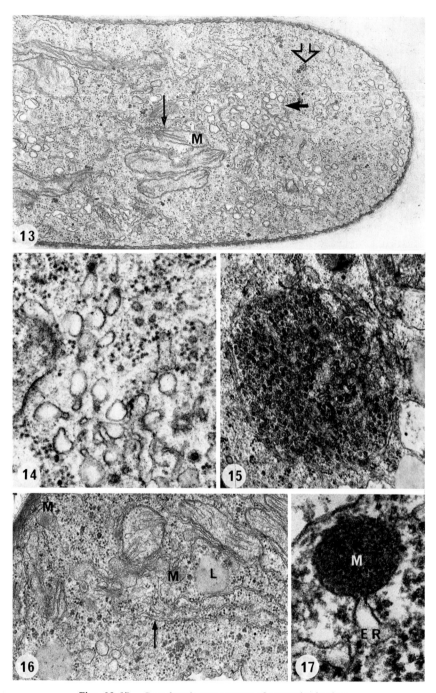

Figs. 13–17. Cytoplasmic components of vegetative hyphae.

fungi and are often associated with one or more cisternae of endoplasmic reticulum (e.g., Fig. 13 and micrographs in Coffey *et al.*, 1972a, and Müller *et al.*, 1974a). The number of cristae, and the density of the surrounding matrix, may vary considerably between different parts of the mycelium. For *Puccinia helianthi* and *Melampsora lini*, Coffey *et al.* (1972a) claim that mitochondria of intercellular hyphae possess fewer cristae than those in haustoria but workers with other rusts have not reported such a phenomenon.

Single membrane bound structures, conforming to the morphological definition of microbodies (Frederick *et al.*, 1968) have been observed in the intercellular mycelium of several rusts. Most commonly they are found associated with the septal pore apparatus (Coffey *et al.*, 1972a; Ehrlich *et al.*, 1968a; Littlefield and Bracker, 1971a; Müller *et al.*, 1974a; Rijo and Sargent, 1974; see p. 118) but they can also be seen dispersed in the cytoplasm (Coffey *et al.*, 1972a; Heath and Heath, 1975). In neither situation, however, are they commonly associated with lipid droplets as found in infection structures (see Chapter 3, Section III,B) but they are often in close association with a cisterna of endoplasmic reticulum (Fig. 2). In axenic cultures of *Melampsora lini*, they seem to develop in clusters, possibly from aggregates of tubular cisternae of endoplasmic reticulum (Fig. 16) (Coffey *et al.*, 1972a), but there are no similar reports concerning intercellular hyphae. Direct connections between microbodies and single cisternae of what may be endoplasmic reticulum have been seen in young infection structures of *Uromyces phaseoli* var. *vignae* (Fig. 17) (Heath and Heath, unpublished), as

---

Fig. 13. Hyphal tip of axenically grown *Melampsora lini* (uredial stage) showing vesicles near the apex and their apparent subapical origin from single tubular cisternae (large closed arrow). Note the parallel platelike arrangement of the cristae in the mitochondria (M) and the association of this organelle with a strand of endoplasmic reticulum (small arrow). The clusters of electron-opaque granules in the cytoplasm (open arrow) may be glycogen. (× 22,000) (From Coffey, 1975.)

Fig. 14. Membranous tubules and vesicles in an intercellular hypha of *M. lini* (uredial stage) which morphologically resemble those thought to have Golgi body function in other septate fungi. (× 37,600) (L. J. Littlefield and C. E. Bracker, unpublished.)

Fig. 15. Tubular complex in an intercellular hypha of *Hemileia vastatrix* (uredial stage). (× 37,600) (From Rijo and Sargent, 1974. Reproduced by permission of the National Research Council of Canada from *Can. J. Bot.* **52**, 1363–1367.)

Fig. 16. Axenically cultured hypha of *M. lini* (uredial stage) showing association of crystal-containing microbodies (M) and long, nonvesicular (compare with Fig. 11) strands of endoplasmic reticulum (arrow). Lipid droplets (L) are also present in this micrograph. (× 34,000) (From Coffey *et al.*, 1972a. Reproduced by permission of the National Research Council of Canada from *Can. J. Bot.* **50**, 231–340.)

Fig. 17. Continuity between the membranes of a microbody (M) and what seems to be endoplasmic reticulum (ER) in a urediospore-derived infection structure of *Uromyces phaseoli* var. *vignae*. (× 110,200) (I. B. Heath and M. C. Heath, unpublished.)

they have in other fungi (Maxwell *et al.*, 1977), but again not in intercellular hyphae. Presumably this lack of information concerning microbodies in established parasitic mycelium reflects the more sparse, scattered, distribution of this organelle (except near the septal pore) in this part of the fungus rather than any difference in ontogeny.

In contrast to the microbodies found in infection structures (see Chapter 3, Section III,B), microbodies in vegetative mycelium often contain crystal lattices with a periodicity of 8 nm (Coffey *et al.*, 1972a) similar to that of catalase-rich crystals in microbodies of other organisms (Frederick and Newcomb, 1969; Vigil, 1973). However, as for infection structure microbodies, attempts at ultrastructural demonstration (DAB test) of catalase activity in the septal pore microbodies of *Puccinia helianthi* (Coffey *et al.*, 1972a) have been unsuccessful. For the significance of this result, see discussion in Chapter 3, Section III,B.

Cytoplasmic microtubules have been reported or illustrated only for intercellular hyphae of *Puccinia helianthi* (Coffey *et al.*, 1972a), *Melampsora lini* (Coffey, 1975) and *Uromyces phaseoli* var. *vignae* (I. B. Heath and M. C. Heath, 1978), but considering the ubiquitous distribution of these structures throughout eukaryotic organisms, it is almost certain that they occur in all rust species. The possible roles of cytoplasmic microtubules in rust fungi are discussed in Section II,A,1,d of this chapter and also Chapter 3, Section III,C, and Chapter 5, Section II.

The two recognizable storage products seen in the cytoplasm of intercellular hyphae are glycogen particles (Fig. 13) and lipid droplets (Fig. 16) (e.g., Coffey *et al.*, 1972a; Van Dyke and Hooker, 1969; Walles, 1974). These vary in distribution and abundance between different studies; for example in *Uromyces phaseoli* var. *vignae* (M. C. Heath, unpublished), glycogen is prevalent only in the older portions of the mycelium but in *Melampsora lini,* glycogen particles can be seen close to the apex (Fig. 13) (Coffey, 1975). Such differences presumably reflect differing physiological activities of these portions of the mycelium but too few detailed studies have been made to determine what general situations are the most common.

A number of authors have commented on the presence of tubular or other types of membrane aggregates in the cytoplasm or vacuole of monokaryotic and dikaryotic intercellular mycelium (Müller *et al.*, 1974a; Pinon *et al.*, 1972; Rijo and Sargent, 1974; Van Dyke and Hooker, 1969; Walles, 1974). Some of these fall under the definition of a lomasome (Moore and McAlear, 1961) but others have been interpreted as Golgi bodies (dictyosomes) (Moore, 1963b; Pinon *et al.*, 1972) or as structures with a similar function (Fig. 15) (Rijo and Sargent, 1974). If these interpretations are correct, the rusts would differ from other septate fungi where the Golgi apparatus is represented by single cisternae *not* aggregated into dictyosomal stacks (Beckett *et al.*, 1974). However, since the apical vesicles of axenically grown *M. lini* (Fig. 13) (and possibly those of germ tubes of *U. phaseoli* var. *vignae*, see Chapter 3, Section III,B, and Fig. 17) seem to arise from single cisternae resembling those thought to have Golgi function in other fungi, it seems likely that the rusts are no different from other septate fungi in their type of Golgi apparatus. Cisternae exhibiting several of the characteristic features of Golgi cisternae in other septate fungi have been seen in nonapical regions of *M. lini* (Fig. 14) (Littlefield and Bracker, unpublished). Possibly the morphological similarity between a single Golgi cisterna and a portion of vesiculated smooth endoplasmic reticulum explains why these cisternae have not been commonly noted in nonapical portions of other rusts. However, it should also be pointed out that the differentiation between cisternae of endoplasmic reticulum and single Golgi cisternae may merely be a matter of semantics since one is most likely derived from the other (Morré, 1975).

Membrane-bound vesicles and vacuoles of various morphologies are common in intercellular hyphae (Fig. 2). As in most fungi, vacuolation is most prominent in older parts of the mycelium; the process of vacuolation, however, has not been studied ultrastructurally in the parasitic mycelium so it remains to be determined whether this differs from that observed in the germ tube (see Chapter 3, Section III,D). Such a difference is possible since, in the germ tube, the absence of any appreciable synthesis of cytoplasm, and the lack of septation, forces vacuole formation to keep pace with tip growth; vacuolation in intercellular mycelium, in contrast, commonly occurs in cytoplasm isolated from the growing tip by septum formation. However, as in the germ tube, no "dense-body" vesicles (see Chapter 3, Section III,D) have been seen in vacuolating portions of intercellular hyphae of *Uromyces phaseoli* var. *vignae* (M. C. Heath, unpublished).

II. INTRACELLULAR STRUCTURES

Like many other biotrophic parasites of higher plants (see reviews by Bracker and Littlefield, 1973; Ehrlich and Ehrlich, 1971b), the rusts characteristically produce "branches" of the intercellular mycelium which penetrate the adjacent host cell walls. The subsequently formed structures are usually referred to as "intracellular" although, as will be shown later, they do not in fact penetrate the host plasmalemma. In infections originating from aeciospores or urediospores, these intracellular structures usually (but not always, see Section II,A, this chapter) show determinate growth and clearly fall within Bushnell's (1972) definition of haustoria. However, recent ultrastructure studies suggest that these haustoria differ in many ways from the intracellular structures of basidiospore-derived infections* and there is some controversy as to whether the latter should be called haustoria or merely intracellular hyphae. This point is discussed later (see Section II,A, this chapter), but for the moment, the term "dikaryotic (D) haustoria" will be used for the intracellular structures produced by aeciospore- and urediospore-derived infections while those formed after basidiospore inoculation will be called "monokaryotic (M) haustoria."† Each will be described separately and the host response to both types of invasion will then be compared. It should be pointed out, however, that there are some structures associated with the haustorium which cannot be ascribed with any certainty to either fungus or host and also the nature of the association between the two organisms makes it impossible to discuss the ultrastructure of one in complete isolation from the other. Thus the authors have used their discretion as to the distribution of certain information under the subheadings which follow.

A. Dikaryotic Haustoria

Of all vegetative structures produced by rust fungi, the D-haustorium has been the one most studied by electron microscopists. Unfortunately, it is also the structure for which there is considerable diversity in terminology in the literature as witnessed by the fact that the authors of this book have, in the past, used different systems. However, rather than erect yet another set of terms, we have chosen to compromise in this volume and use the terminology suggested by Bushnell (1972) with the exceptions that "papilla" has been replaced by the more generally used "collar" (if it occurs around the haustorial neck) and that "sheath" has been replaced by "encasement" (Bracker and Littlefield, 1973). Our reason for the latter substitution is that, in our opinion, much of the confusion in haustorium terminology results from the varied use of the word "sheath." Thus, instead of arguing for the precedence of one usage over another, it seems more sensible to eliminate the word altogether. A diagram depicting the names and spatial relationships of the structures associated with the D-haustorium is shown in Fig. 18. This terminology is followed in this book regardless of that used by the

---

*Studies of microcyclic rusts (e.g. *Puccinia malvacearum,* R. F. Allen, 1935) suggest that it is the type of spore used to *initiate* the infection which determines the type of haustorium subsequently produced; the type of spore ultimately *produced* by the infection seems irrelevant in this respect.

†We have chosen this term, not to imply information concerning nuclear condition, but rather to emphasize the point made in the preceding footnote; the terms "pycnial" or "aecial" haustoria cannot be used in microcyclic rusts where pycnia and aecia do not exist.

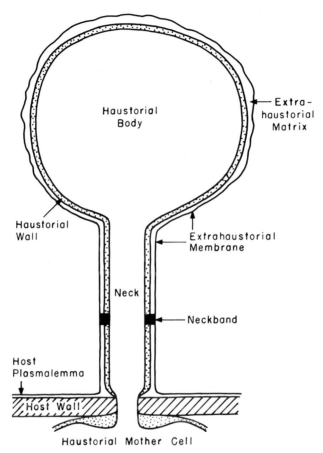

Fig. 18. Diagrammatic representation of the fungal and host structures (and terminology) associated with the typical dikaryotic haustorium.

authors cited. Figure 19 is a diagrammatic representation of the events accompanying haustorium formation. It will be discussed further in Section II,A,2. Figure 20 presents the mode of development of collars in rust infections and is discussed in Section II,C,2

1. Mature Haustoria

a. The haustorial mother cell. A survey of studies using light microscopy (e.g., Rice, 1927 and references therein) and scanning electron microscopy (Gold *et al.,* 1979; Kinden and Brown, 1975a; Plotnikova *et al.,* 1979; Pring and Richmond, 1975) shows that the dikaryotic haustorium typically consists of a relatively large body connected by a tubular neck (Fig. 21) to a terminal portion of the intercellular mycelium delimited by a septum. This latter por-

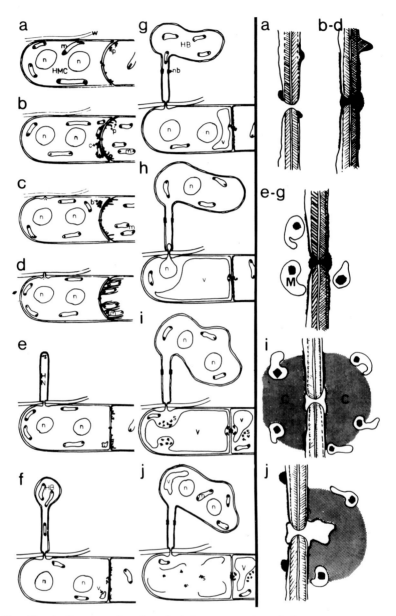

Fig. 19. A diagrammatic, chronological representation of the events accompanying haustorium formation (left) in *Uromyces phaseoli* var. *vignae,* and the correlated state of the septal pore apparatus of the haustorial mother cell septum (right). On the left, b, multivesicular body; c, irregular plasmalemma elaborations; HB, haustorial body; HMC, haustorial mother cell; HN, haustorial neck; m, mitochondrion; n, nucleus; nb, neckband; p, plasmalemma protrusions; s, glycogen-like granules; v, vacuole; w, host cell wall. On the right, C, differentiated region of cytoplasm; M, microbody. In all cases the haustorial mother cell is to the left of the septum. (From Heath and Heath, 1975.)

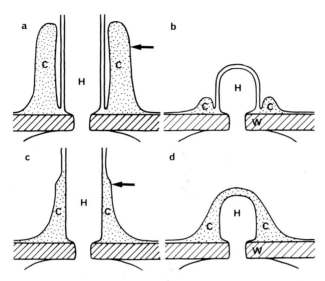

Fig. 20. Diagrammatic representation of the possible mode of development and final appearance of the two types of collars found in rust infections. (a) Mature, Type I collar, C. Note how the host plasmalemma (arrow) lines the inside of the collar and folds back to run along the haustorial neck, H. (b) Possible mode of formation of the Type I collar involving the "growing up" of collar material, C, from the host cell wall, W, after the fungus, H, has entered the cell. (c) Mature, Type II collar, C. Note how the plasmalemma (arrow) does *not* fold back along the inside of the collar but continues directly along the haustorial neck, H. (d) Possible mode of formation of a Type II collar:collar material, C, is deposited either on the host cell wall, W, before penetration or directly onto the fungus, H, soon after penetration. If the fungus breaks through this material, the latter will be left as a collar around the haustorial neck.

tion of the haustorial apparatus is known as the haustorial mother cell (HMC) and appears closely appressed to the host wall (e.g., Littlefield, 1974) in the region of the penetration site of the haustorium. Transmission electron microscopy shows that an amorphous, electron-opaque layer, probably continuous with that covering the rest of the intercellular mycelium is often seen on the outside of the HMC in this region and commonly fills part of the angle of contact between host and fungus (Fig. 22) (Abu-Zinada *et al.*, 1975; Coffey, 1976; Coffey *et al.*, 1972a; Ehrlich *et al.*, 1968b; Hardwick *et al.*, 1971; Heath, 1972; Heath and Heath, 1971; Kajiwara, 1971; Littlefield and Bracker, 1972; Müller *et al.*, 1974a; Oláh *et al.*, 1971; Rijkenberg and Truter, 1973a; Van Dyke and Hooker, 1969). Most likely this material serves to stick the HMC to the host wall (Hardwick *et al.*, 1971; Littlefield and Bracker, 1972; Müller *et al.*, 1974a; Rijkenberg and Truter, 1973a), giving it a similar role to that suggested in other regions of the mycelium. Most workers have assumed this material to be of fungal origin. However, it is also possible that there may be a host component present since Mendgen and

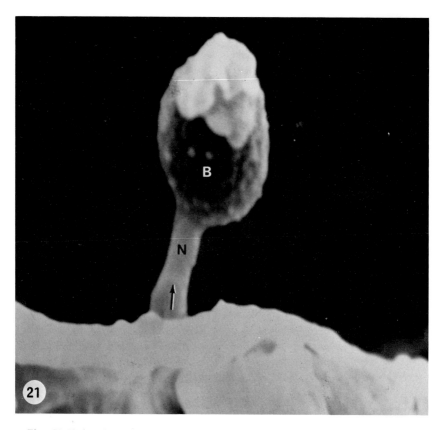

Figs. 21-27 (continued).   Ultrastructure of the mature dikaryotic (D) haustorium.

Fig. 21.   Scanning electron micrograph of a D-haustorium of *Puccinia graminis* f. sp. *tritici.* Note the clear differentiation between the body (B) and the neck (N), and the presence of a ring (arrow) around the latter which may correspond to the neckband shown in Fig. 22. Remnants of host cytoplasm can be seen draped on top of the haustorial body. (× 14,000) (Fixed in glutaraldehyde; Y. A. Plotnikova and L. J. Littlefield, unpublished.)

Fuchs (1973) have shown ultrastructurally that it has peroxidase activity. Presumably, this enzyme is of host origin since similar activity could not be demonstrated anywhere within the fungal protoplast, but was found in other situations within the higher plant.

At the site of host penetration, the HMC wall is usually thickened (Figs. 22, 26, 42, 43) (Abu-Zinada *et al.*, 1975; Coffey, 1976; Coffey *et al.*, 1972a; Ehrlich and Ehrlich, 1963; Harder, 1978; Hardwick *et al.*, 1971; Heath and Heath, 1971, 1975; Kajiwara, 1971; Littlefield and Bracker, 1972; Manocha and Shaw, 1967; Müller *et al.*, 1974a; Rijkenberg and Truter, 1973a; Rijo and Sargent, 1974; Shaw and Manocha, 1965b). In *Melampsora lini*

(Littlefield and Bracker, 1972) and *Uromyces phaseoli* var. *vignae* (Heath and Heath, 1971), the thickening has been attributed to the presence of an electron-opaque, lens-shaped layer deposited *between* the two fibrillar layers (i.e., excluding the amorphous outer layer) which comprise the rest of the HMC wall (Figs. 26 and 42). Some micrographs of *U. phaseoli* var. *typica* (Hardwick *et al.*, 1971), however, show the HMC wall to be composed of three or more layers and the thicker region around the penetration site to be continuous with the outer fibrillar layer. Yet another variation can be seen in *Hemileia vastatrix* (Rijkenberg and Truter, 1973a; Rijo and Sargent, 1974) where the thicker region contains one or more additional layers deposited against the inner surface of the HMC wall. The reality of these differences, however, is questioned by the fact that different interpretations can be made from micrographs produced by different workers of the same rust species, for example, *M. lini* (Figs. 22 and 28) (Littlefield and Bracker, 1972 versus Coffey *et al.*, 1972a) and *U. phaseoli* var. *typica* (Hardwick *et al.*, 1971 versus Müller *et al.*, 1974a). In addition, a closer look at a selection of penetration sites of *U. phaseoli* var. *vignae* (M. C. Heath, unpublished) reveals some where the thicker region of the HMC wall appears homogeneous and unlayered, some where there appears to be an interpolation of a middle wall layer (Fig. 42) and some where an additional inner layer seems to have been added (Fig. 43). These results suggest that the appearance and position of wall "layers" in the penetration region of the HMC are highly sensitive to minor differences in preparative procedures. Obviously more information is needed to determine exactly the nature of the changes in the HMC wall associated with haustorium formation.

When attached to a mature haustorium, the haustorial mother cell is vacuolate and is lined with only a thin layer of cytoplasm continuous with that of the haustorium (Fig. 22). The normal cytoplasmic components are present (see Section I, this chapter) but nuclei are absent since they migrate with the majority of the cytoplasm into the haustorium during its formation (see Haustorium Development, Section II,A,2, this chapter). In *Melampsora lini* (Coffey *et al.*, 1972a), *Puccinia sorghi* (M. C. Heath, unpublished), and *Uromyces phaseoli* var. *vignae* (Heath and Heath, 1975), glycogen seems to accumulate in the cytoplasm of haustorial mother cells of older haustoria (Fig. 22) and, in the latter fungus, often appears to be released into the vacuole following tonoplast disintegration.

b. The penetration peg. The penetration peg which breaches the host wall arises from the center of the thickened region of the haustorial mother cell (HMC). In most studies, the development of the peg does not seem to be restricted to any special portion of the host cell but in telial galls of *Gymnosporangium juniperi-virginianae*, the pegs seem to be located only at the

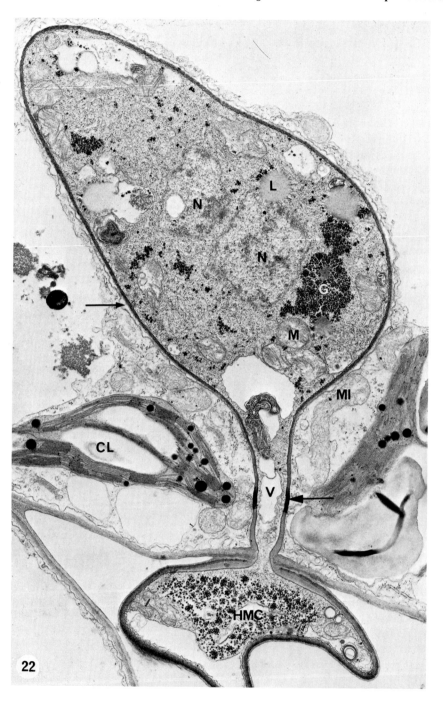

site of pits in the otherwise thick host cell wall (Mims and Glidewell, 1978). Details of peg formation are discussed under Haustorium Development (Section II,A,2, this chapter) but in the mature haustorium, the diameter of the peg is usually less than 0.5 μm and the fungal wall in this region is extremely thin (Fig. 24) and often cannot be clearly delimited from the host wall (Fig. 23). Nevertheless, claims that the fungal wall is absent (Bossányi and Oláh, 1974; Calonge, 1969; Ehrlich and Ehrlich, 1971b) must be viewed with some caution when one considers the considerable difference in appearance and clarity of fungal and host walls in micrographs of the same rust prepared by different workers (e.g., *Melampsora lini*, Coffey *et al.*, 1972a versus Littlefield and Bracker, 1972). Assuming that a wall, albeit very thin, is indeed present in the penetration peg, then the question arises as to which layer of the HMC wall it is connected to. No clear answer can be made, however, since conventional staining usually suggests it to be the inner layer (Figs. 22 and 23) (Coffey *et al.*, 1972a; Heath and Heath, 1975; Littlefield and Bracker, 1972; Shaw and Manocha, 1965b; see also Haustorium Development, Section II,A,2, this chapter) while treatment of *M. lini* sections with periodic acid-chromic acid-phosphotungstic acid (PACP stain) results in the wall of the penetration peg staining similarly to the middle layer of the HMC wall (Fig. 26). However, in all studies and staining procedures the outer, fibrillar wall layer of the HMC does *not* appear to be continuous with the wall of the peg (Fig. 23). All studies also seem to agree that penetration of the host wall must involve enzymatic action rather than physical force since the host wall, and often the wall fibrils, end abruptly at the penetration peg and do not appear distorted or displaced inward (Figs. 23 and 24) (Bossányi and Oláh, 1974; Coffey *et al.*, 1972a; Ehrlich and Ehrlich, 1963; Hardwick *et al.*, 1971; Heath and Heath, 1971; Kajiwara, 1971; Littlefield and Bracker, 1972). However a bulging *outward* of the host wall has been reported for *Hemileia vastatrix* infections (Rijo and Sargent, 1974); this phenomenon is discussed further under Host Responses (Section II,C, this chapter).

c. The haustorial neck. On the inner side of the host wall, the fungus expands to form a tubular neck about twice the width of the penetration peg (Fig. 22). Many early light microscopists suggested that the haustorium in-

Fig. 22. Transmission electron micrograph of a haustorium of *Melampsora lini*. The haustorial body contains two nuclei (N), mitochondria (M), lipid droplets (L), glycogen (G), and many ribosomes. The haustorial neck is vacuolate (V), bears a conspicuous, electron-opaque neckband (large arrow), and is connected to the collapsed, vacuolate, haustorial mother cell (HMC) outside the host cell. Note the presence of host chloroplasts (CL) and mitochondria (MI) adjacent to the haustorium and the electron-lucent extrahaustorial matrix (small arrow) around the haustorial body. ($\times$ 16,000) (From Coffey *et al.*, 1972a. Reproduced by permission of the National Research Council of Canada from *Can. J. Bot.* **50**, 231–240.)

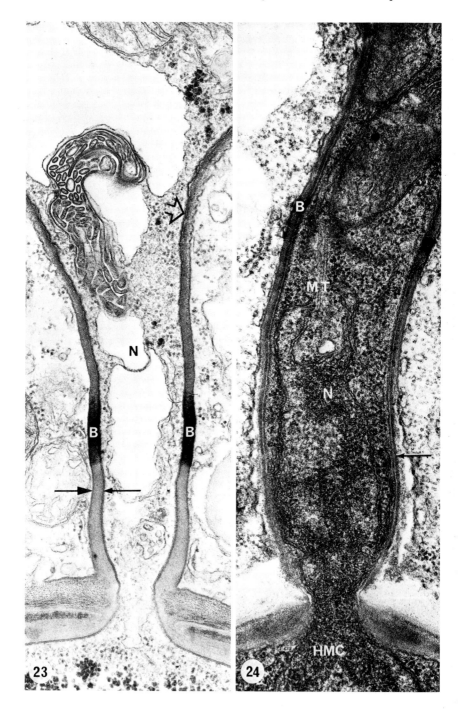

vaginated, rather than punctured, the host protoplast (Rice, 1927 and references therein; Thatcher, 1939) but it took electron microscopy to prove this conclusively. Today there is abundant ultrastructural evidence to show that young (Fig. 32) and mature haustoria (Fig. 23) are separated from the host cytoplasm by an "extrahaustorial membrane" (see Fig. 18). However, indisputable connection of this membrane with the host plasmalemma at the base of the haustorial neck has proved to be more difficult to demonstrate due to the convolutions often present in this region. Nevertheless, such a connection has been shown for several rust infections (Fig. 23) (Coffey *et al.*, 1972a; Ehrlich and Ehrlich, 1971b; Hardwick *et al.*, 1971; Littlefield and Bracker, 1970) and there is no ultrastructural evidence to suggest the existence of any alternative situation.

In all rust species which have been examined for this feature, the region between host and fungal protoplast is bridged by electron-opaque material about half way along the haustorial neck. This neckband (or neck ring) (see Fig. 18) can either appear more or less uniform in density (Fig. 23) (Coffey *et al.*, 1972a; Littlefield and Bracker, 1972), or may appear more diffuse at its upper and/or lower boundaries (Fig. 24) (Hardwick *et al.*, 1971; Heath and Heath, 1971; Rijo and Sargent, 1974). Figure 24 shows clearly that the neckband fills the region between the extrahaustorial membrane and the fungal plasmalemma and also seems to encompass and extend along these membranes. A rather unusual situation has been shown by Littlefield and Bracker (1972) where an oblique section of the neckband of *Melampsora lini* contained membrane-bound electron-opaque circles which were interpreted as projections of the neckband into the lumen of the neck (Fig. 27). Such projections must be very short or rare, however, since they could not be seen in longitudinal sections and they have not been reported for other rusts. The composition of the neckband is unknown but it usually stains intensely with

Fig. 23. The neck (N) of the same haustorium of *Melampsora lini* shown in Fig. 22. Note the uniform staining of the region between the fungal plasmalemma (small solid arrow) and the invaginated host plasmalemma (extrahaustorial membrane) (large solid arrow) on either side of the neckband (B). Only at the junction of the neck with the body (open arrow) does an outer, more electron-lucent, extrahaustorial matrix become apparent. (× 54,000) (From Coffey *et al.*, 1972a. Reproduced by permission of the National Research Council of Canada from *Can. J. Bot.* **50**, 231–240.)

Fig. 24. The neck of a young D-haustorium of *Uromyces phaseoli* var. *vignae.* One of the two nuclei (N) can be seen apparently migrating from the haustorial mother cell (HMC) into the haustorium. Microtubules (MT) are always seen in the haustorial neck at this time in this species. Note the narrower neckband (B) compared with that in Fig. 23 and the clear distinction of two layers between the fungal and host plasmalemmas, both above and below the neckband. The inner, electron-opaque layer (arrow) seems to be fungal wall since it is continuous with the wall of the haustorial mother cell. The outer layer is interpreted as extrahaustorial matrix of uncertain origin. (× 46,900) [From Heath and Heath, 1971. Reproduced by permission from *Physiol. Plant Pathol* **1**, 277–287. Copyright by Academic Press, Inc. (London) Ltd.]

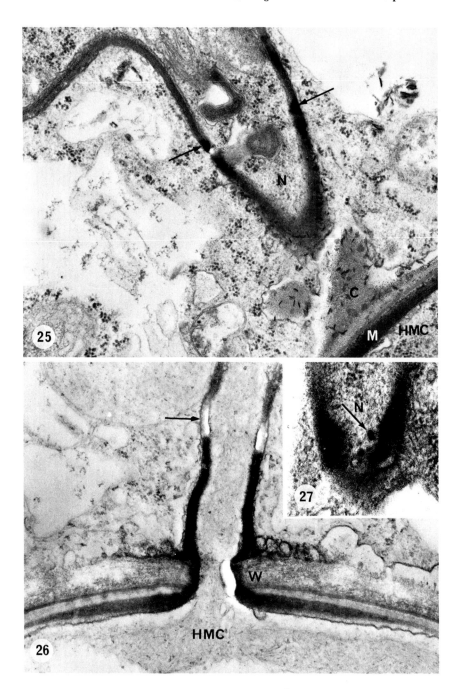

osmium tetroxide (Fig. 23) (Coffey, 1976; Coffey *et al.*, 1972a; Harder, 1978; Hardwick *et al.*, 1971; Heath, 1972; Heath and Heath, 1971; Littlefield and Bracker, 1972; Mims and Glidewell, 1978; Müller *et al.*, 1974a; Rijkenberg and Truter, 1973a; Rijo and Sargent, 1974; Van Dyke and Hooker, 1969) and potassium permanganate (Ehrlich and Ehrlich, 1971b; Hardwick *et al.*, 1971) and is preferentially removed by periodic acid (Fig. 26) (Littlefield and Bracker, 1972). Bossányi and Oláh (1974) suggest that in *Puccinia graminis* f. sp. *tritici*, the neckband is associated with acid phosphatase activity but since they used osmium tetroxide to fix their material, it is not clear whether the observed electron opacity is due to enzyme activity or the normal osmiophilic nature of this region. A structure which may be the neckband can also sometimes be seen in glutaraldehyde-fixed, entire haustoria examined by scanning electron microscopy (Fig. 21).

The tight association of the host and fungal membranes with the neckband is reminiscent of the Casparian strip of endodermal cells of vascular plants (Littlefield and Bracker, 1972; Heath, 1976b) and, in *Puccinia sorghi*, it does indeed seem to similarly prevent the apoplastic flow of uranyl acetate (and presumably other solutes) along the neck (Fig. 25) (Heath, 1976b). Such a restriction to apoplastic "escape" of host solutes released into the extrahaustorial matrix would seem to favor the uptake of substances into the body of the haustorium and one might expect the equivalent of a neckband to be advantageous to any haustorial parasite (Heath, 1976b). It is interesting, therefore, that a morphologically similar neckband has been demonstrated around some powdery mildew haustoria (Gil and Gay, 1977) but has not yet been reported for haustoria of other fungal groups. Bossányi and Oláh (1974) suggest for *Puccinia graminis* f. sp. *tritici* that the neckband

---

Fig. 25. D-haustorium of *Puccinia sorghi* fixed after the infected tissue had been allowed to transpire aqueous uranyl acetate for 24 hours. The haustorial neck (N) was connected, out of the plane of section, to the haustorial mother cell (HMC) lying in the intercellular space. Typical, needle-shaped uranyl crystals can be seen in the wall of the haustorial mother cell (M), in the collar around the base of the haustorial neck (C) and in the region between the plasmalemmas of host and parasite up to, but not beyond, the neckband (arrows). (× 32,800) (From Heath, 1976b. Reproduced by permission of the National Research Council of Canada from *Can. J. Bot.* 52, 2591–2597.)

Fig. 26. A section of a penetration site of a D-haustorium of *Melampsora lini* treated with periodic acid–chromic acid–phosphotungstic acid (PACP stain). The neckband (arrow) appears to have been etched away while the neck wall below the neckband stains strongly and seems continuous, through the host wall (W), with the middle layer of the wall of the haustorial mother cell (HMC). The haustorial wall distal to the neckband stains less strongly. (× 37,600) (From Littlefield and Bracker, 1972.)

Fig. 27. Oblique section through a neckband of *M. lini* conventionally fixed and stained. The electron-opaque circular areas (arrow) are interpreted as evaginations of the neckband into the lumen of the haustorial neck (N). (× 60,000) (From Littlefield and Bracker, 1972.)

marks a site of exchange between the fungus and its host but such a hypothesis is not supported by clear ultrastructural evidence. However the neckband of *Melampsora lini* does seem to mark a change in composition of the material between fungus and host protoplasts (see below, this section) and also in the stainability of the extrahaustorial membrane with the periodic acid–chromic acid–phosphotungstic acid (PACP) stain (Littlefield and Bracker, 1972; see also Host Responses, Section II,C, 1).

Although neckbands have been observed in all rusts where they have been looked for in D-haustoria, among individual haustoria some are more clearly differentiated than others. This may account for the fact that they have been reported as occasionally absent in *Uromyces fabae* (Abu-Zinada *et al.*, 1975) (or the investigators may have been observing young haustoria before the neckband had developed, see Section II,A,2, this chapter). The reason for this variability in staining is unknown but in the authors' experience, a tight association of adjacent membranes can be found even in the absence of strong staining, suggesting a functional neckband is still present.

In contrast to the constancy of the neckband among species, micrographs of haustorial necks seem to show some variation in the other structures found between the protoplasts of fungus and host. At one extreme (Fig. 23) the region proximal to the neckband (i.e., toward the HMC) stains uniformly and has been described as consisting entirely of fungal wall (*Melampsora lini*, Coffey, 1976; Coffey *et al.*, 1972a; Littlefield and Bracker, 1972; *Puccinia coronata* f. sp. *avenae*, Harder, 1978) and consistent with this interpretation is the strong, uniform, staining of this region in *M. lini* infections after PACP staining (Fig. 26) (Littlefield and Bracker, 1972). In addition, the host plasmalemma seems tightly bound to the "fungal wall" of *M. lini* since it is not displaced after plasmolysis of the host cell (Allen *et al.*, 1979). On the distal side of the neckband, however, the homogeneous "wall" gives rise to two layers, an inner more electron-opaque layer and an outer more electron-lucent region, both of which also surround the haustorial body (Fig. 23) (Coffey, 1976; Coffey *et al.*, 1972a). Coffey has interpreted the outer layer as being extrahaustorial matrix and the inner as being a new type of fungal wall. That the fungal wall of *M. lini* does indeed change in some way at the neckband is also indicated by the fact that the region between the fungal plasmalemma and the extrahaustorial membrane distal to the neckband does not stain as strongly with the PACP stain as that on the other side (Fig. 26).

At the other extreme, micrographs of *Uromyces phaseoli* var. *typica* (Hardwick *et al.*, 1971; Müller *et al.*, 1974a), *U. phaseoli* var. *vignae* (Heath and Heath, 1971), *Puccinia helianthi* (Fig. 5 of Coffey *et al.*, 1972a) and *P. sorghi* (M. C. Heath, unpublished) all clearly show *two* distinct layers proximal to the neckband (Fig. 24). In all but *P. helianthi* (where the penetration region is not illustrated), the inner layer appears to be continuous with the wall of the penetration peg, and is therefore presumably of fungal origin, and the outer shows varying degrees of electron opacity and has been interpreted as extrahaustorial matrix (Heath and Heath, 1971) or as a rudimentary collar (see Host Responses, Section II,C, this chapter) of host origin (Hardwick *et al.*, 1971; Müller *et al.*, 1974a). As found in the *Melampsora lini* type of haustorium, two layers can also be distinguished distal to the neckband; the inner is continuous with, and of equal staining and thickness to, the fungal wall proximal to the neckband and the outer becomes gradually more electron-lucent (Fig. 24). These layers continue around the haustorial body and have been interpreted as fungal wall and extrahaustorial matrix, respectively (Heath and Heath, 1971), or have not been interpreted at all (Hardwick *et al.*, 1971). Support for the suggestion that two layers are present along the *whole* neck is provided in *U. phaseoli* var. *vignae* by the very clear distinction of these layers before the neckband develops (Fig. 32) (Heath and Heath, 1975). That only one of these is fungal wall is suggested by the observation in *U. phaseoli* var. *typica* that when the extrahaustorial membrane fragments in certain incompatible plants, the layer interpreted as matrix material disperses both above and below the neckband leaving the layer interpreted as fungal wall intact (see Chapter 6, Fig. 37). Furthermore, the presence of material other than of fungal origin proximal to the neckband in this latter rust is indicated by the ultrastructural localization of peroxidase activity in this region (Fig. 36); such activity is most likely of host origin since it has not been demonstrated elsewhere in the fungal protoplast but is found in host structures (Mendgen, 1975; Mendgen and Fuchs, 1973).

Thus, there is ultrastructural evidence of two types of organization of the neck region; one where the neck (and body) is covered by a wall of uniform thickness which is *always* separated from the extrahaustorial membrane by material (which may or may not contain collar material, see Section II,C,2, this chapter) distinct from the fungal wall, and the other where the fungal wall is thicker below the neckband and where an extrahaustorial matrix occurs only around the haustorial body and upper part of the neck. Whether this difference is a real one remains to be determined. It is conceivable that the apparent absence of an extrahaustorial matrix below the neckband is an artifact resulting from both wall and matrix being composed of, or permeated by, materials of similar ultrastructural staining characteristics. Alternatively, the extrahaustorial matrix around *any* region of the haustorium may be an artifact of preparation (see following section, this chapter).

The cytoplasm in the neck of the mature haustorium is continuous with that remaining in the haustorial mother cell (HMC) (Fig. 22) and, in general, contains no distinctive features when compared with the cytoplasm of other parts of the fungus. Coffey *et al.* (1972a) report that glycogen is normally absent from the neck of *Melampsora lini,* although present in abundance in the HMC and haustorial body, but such a phenomenon has not been reported for other rusts. A more consistent feature of several species, however, is the presence of an aggregation of concentric, coiled, or tubular membranes in the neck (Fig. 23) (*M. lini,* Coffey *et al.,* 1972a; *Uromyces fabae,* Abu-Zinada *et al.,* 1975; *U. phaseoli* var. *typica,* Hardwick *et al.,* 1971; Müller *et al.,* 1974a; *U. phaseoli* var. *vignae,* Heath and Heath, 1971). The significance of this observation is unknown, but for undisclosed reasons Coffey *et al.* (1972a) regard this structure as a fixation artifact.

d. The haustorial body. In D-haustoria, the haustorial neck terminates in a relatively large body (Figs. 21 and 22) whose shape may vary depending on the rust species (see light microscope study of Rajendren, 1972). Around this body, the protoplasts of fungus and host are normally separated by two or more ultrastructurally distinct layers and where two layers are present (Figs. 22 and 37), most workers interpret the more electron-opaque layer next to the fungal plasmalemma as fungal wall while the remaining more electron-lucent layer is considered part of the extrahaustorial matrix (Calonge, 1969; Harder, 1978; Heath, 1972; Mims and Glidewell, 1978; Pinon *et al.,* 1972; Shaw and Manocha, 1965b; Van Dyke and Hooker, 1969). However, where more than one electron-opaque layer is present (Figs. 38 an 68), some workers have considered all but the innermost one to be extrahaustorial matrix (Müller *et al.,* 1974a; Rijkenberg and Truter, 1973a; Rijo and Sargent, 1974; Shaw and Manocha, 1965b; Zimmer, 1970) while others have considered all to be part of the fungal wall (Coffey *et al.,* 1972a; Ehrlich and Ehrlich, 1963; Heath and Heath, 1971; Kajiwara, 1971). Since there is ultrastructural evidence that at least some electron-opaque material in the extrahaustorial matrix is of host origin (see Host Responses, Section II,C, this chapter), it seems likely that in most cases the former interpretation is correct. Indeed, in a later paper, Coffey (1976) changes his mind and interprets as extrahaustorial matrix the material originally interpreted as fragmenting fungal wall (Coffey *et al.,* 1972a). However, in *U. phaseoli* var. *vignae* (Heath and Heath, 1971) and *Puccinia graminis* f. sp. *tritici* (Ehrlich and Ehrlich, 1963), the so-called second layer of the haustorial wall is much more discrete and uniform in thickness than the material ascribed by others to the extrahaustorial matrix (compare Figs. 38 and 68) and as yet there is no good morphological reason to suggest that it is part of the latter. It is possible, however, that it may represent a region of the fungal wall impregnated by matrix material, rather than a distinct wall layer.

Accepting the interpretations of the authors mentioned above, the extrahaustorial matrix around the haustorial body can be almost entirely electron-lucent (Fig. 22) (Abu-Zinada *et al.,* 1975; Coffey *et al.,* 1972a;

Harder, 1978; Heath, 1972; Heath and Heath, 1971; Littlefield and Bracker, 1972; Van Dyke and Hooker, 1969) or uniformly granular (Calonge, 1969; Mims and Glidewell, 1978). Alternatively it may contain irregular or more discrete layers of amorphous or fibrillar electron-opaque material (Fig. 68) (Coffey, 1976; Coffey et al., 1972a; Ehrlich and Ehrlich, 1971b; Harder, 1978; Kajiwara, 1971; Manocha and Shaw, 1967; Müller et al., 1974a; Orcival, 1969; Rijkenberg and Truter, 1973a; Rijo and Sargent, 1974; Shaw and Manocha, 1965b; Zimmer, 1970). In Hemileia vastatrix, oblique sections through this electron-opaque material suggest that it is composed of hollow fibrillar elements about 23 nm in diameter but a similar morphology has not been reported for matrix material in other rusts. Several workers, however, claim that the electron-opaque component increases with the age of the haustorium (Coffey et al., 1972a; Ehrlich and Ehrlich, 1971b; Harder, 1978; Manocha, 1975; Manocha and Shaw, 1967; Orcival, 1969; Shaw and Manocha, 1965b; Zimmer, 1970) and whether this relates to the occasional detection of peroxidase activity in the inner portion of the matrix of Uromyces phaseoli var. typica (Mendgen and Fuchs, 1973) remains to be determined. The overall thickness of the matrix has also been claimed to increase with haustorium age (Ehrlich and Ehrlich, 1971b; Harder, 1978; Manocha and Shaw, 1967; Oláh et al., 1971; Orcival, 1969; Shaw and Manocha, 1965b) but Kajiwara (1971) has reported the reverse. The validity of these reports is difficult to judge, however, since the apparent thickness of the matrix depends on the plane of section relative to the curvature of the haustorium and none of the above-mentioned studies provides evidence that comparisons were made from sections known to be taken at right angles to the fungal wall. It could be that the "increase" in matrix thickness is primarily due to the more irregular shape of the older haustorial body making glancing sections more common and more difficult to recognize.

One final point which should be considered in relation to the extrahaustorial matrix is whether it is an artifact of preparation — a proposal which can be made with some justification when this region appears completely or irregularly electron-lucent. As yet this question cannot be answered conclusively and the best argument for the reality of this region is the fact that it can be found around all rust haustoria (including those fixed with glutaraldehyde and then freeze-etched, Littlefield and Bracker, 1972) as well as around intracellular structures in a wide variety of symbiotic relationships (Bracker and Littlefield, 1973) fixed and prepared by a variety of procedures. Nevertheless, there is a strong possibility that the electron-lucent "matrix" arises through loss of turgor during fixation, allowing the fungal wall to shrink from the adpressed extrahaustorial membrane in regions where the two structures are not firmly attached (F. H. E. Allen et al., 1979). Such an artifactual separation of wall and membrane would explain why, when the matrix contains electron-opaque components, this stained material often looks rather frayed where it abutts the surrounding electron-lucent region or seems to have separated into two layers, one portion against the fungal wall and the other against the extrahaustorial membrane. This concept of the electron-lucent matrix as an artifact of preparation is supported by light microscope observations of the powdery mildew, *Erysiphe graminis,* where the extrahaustorial matrix was seen to expand when chemical fixatives were added (Bushnell, 1972). Bushnell (1972) concludes from observations of undisturbed, living, haustoria that the extrahaustorial membrane is in close contact with the young, functional, powdery mildew haustorium. If it turns out that the electron-lucent matrix around rust haustoria is not an artifact, then its dispersion once the extrahaustorial membrane has broken (see Chapter 6, Fig. 19) (Heath, 1972; Heath and Heath, 1971; Van Dyke and Hooker, 1969) suggests it to be basically fluid in nature. Its composition, however, is a matter of speculation. Whether electron-lucent or -opaque, there is no good ultrastructural evidence that the matrix contains material secreted by the fungus. There is, however, some indication that the region contains substances of host origin and this is discussed further under Host Responses (Section II,C,2, this chapter). The role of the matrix in incompatible interactions is discussed in Chapter 6.

The fungal cytoplasm of the haustorial body is continuous with that of the neck and HMC (Fig. 22) and, like these regions, seems to have no structural specialization when compared to intercellular hyphae. Two nuclei are usually present (Fig. 22) (see Chapter 5, Section I, for a description of their appearance) (Abu-Zinada *et al.,* 1975; Coffey *et al.,* 1972a; Hardwick *et al.,* 1971; Heath and Heath, 1975; Littlefield and Bracker, 1972; Manocha and Shaw, 1967; Mendgen, 1973a; Rijkenberg and Truter, 1973a) as are profiles of endoplasmic reticulum and mitochondria. Coffey *et al.* (1972a) report (but with no quantitative data) that mitochondrial cristae of *Melampsora lini* and *Puccinia helianthi* are more numerous in the haustorium than in intercellular hyphae and that in the former these organelles more frequently contain whorled aggregations of membranes. However, although Manocha and Shaw (1967) report haustorial mitochondria of *M. lini* to be more elongated than those in germ tubes, differences in cristae are not mentioned. Similar reports have not been made for other rusts. Cytoplasmic microtubules have been illustrated in the haustorial bodies of *P. helianthi* (Coffey *et al.,* 1972a) and *Uromyces phaseoli* var. *typica* (Hardwick *et al.,* 1971) and have been suggested to be involved in maintaining the asymmetric shape of the mature haustorium (Coffey *et al.,* 1972a). However, the rusts seem capable, like other fungi, of maintaining their shape in the absence of any cytoplasm (e.g., infection structures, see Chapter 3, Section III,A) so that this suggestion implies an unusual lack of strength or rigidity of the haustorial body wall which has yet to be demonstrated. Lomasomes or multivesicular bodies are occasionally seen in the haustorial body (Bossányi and Oláh, 1972; Coffey et al., 1972a; Hardwick et al., 1971; Kajiwara, 1971; Shaw and Manocha, 1965b; Van Dyke and Hooker, 1969; Zimmer, 1970) and in spite of suggestions to the contrary, there is no ultrastructural evidence to suggest these structures have an important role in host–parasite interactions (Bossányi and Oláh, 1972) or fungal wall synthesis (Zimmer, 1970). For further discussion of the role of lomasomes in fungi, the reader is referred to Bracker (1967) and Heath and Greenwood (1970). Glycogen (Coffey et al., 1972a; Van Dyke and Hooker, 1969; Zimmer, 1970), lipid bodies (Coffey et al., 1972a; Pinon et al., 1972; Rijo and Sargent, 1974; Van Dyke and Hooker, 1969), and vacuoles (Abu-Zinada et al., 1975; Heath and Heath, 1975; Pinon et al., 1972; Rijo and Sargent, 1974) have all been reported in haustoria (Fig. 22) and may increase in number or size as the haustorium ages (Heath and Heath, 1975). However, as pointed out by Mendgen (1973a), microbodies are scarce in haustoria of Uromyces phaseoli var. typica and, presumably, other rusts since they are not even mentioned in other studies of D-haustoria.

Thus, in general, the ultrastructure of the haustorial cytoplasm supports the histochemical evidence at the light microscope level (Tschen and Fuchs, 1968; Whitney et al., 1962) that the haustorium has no specialized activity which distinguishes it from other portions of the vegetative mycelium. However, it should be mentioned that not all physiological differences may be reflected by differences in ultrastructure and we are also a long way from being able to accurately interpret metabolic activity from electron micrographs. For example, there are reports that draw attention to the fact that most of the numerous ribosomes seen in the haustorial cytoplasm (Fig. 22) appear not to be associated with endoplasmic reticulum (Coffey et al., 1972a; Littlefield and Bracker, 1972) or that typical polysomes are rare (Hardwick et al., 1971). As Coffey et al. (1972a) point out, the abundance of ribosomes in the young haustorium makes the lack of such associations or configurations difficult to distinguish with certainty but similar observations have been made in older haustoria where the number of ribosomes per unit volume seems to be lower (M. C. Heath, unpublished). What all of these observations mean (if anything) in terms of degree, or type, of protein synthesis, however, is unknown.

Although continuity between the cytoplasm of the host and the haustorial body has been reported for several rusts (Calonge, 1969; Ehrlich and Ehrlich, 1963, 1971b), only one of these claims is supported by a clear electron micrograph (Ehrlich and Ehrlich, 1971b) and even in this case, the continuity of membranes is not as convincing as one would wish. No other published studies on rust haustoria report the presence of such cytoplasmic channels in

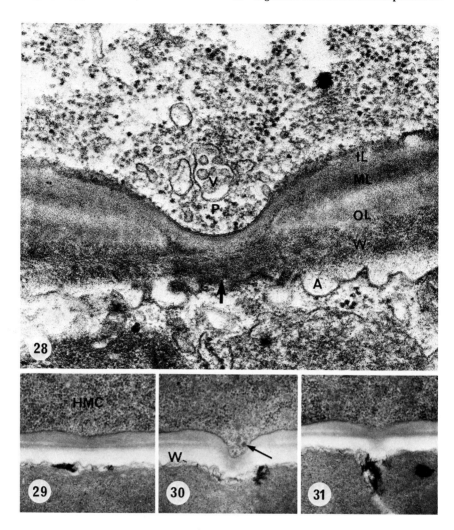

Figs. 28–38.   Development of the dikaryotic (D) haustorium.

Fig. 28.   A young penetration peg (P) of *Melampsora lini.* The inner layer of the fungal wall (IL) has evaginated through the outer layer (OL) and is in contact with the host cell wall (W). The fate of the middle fungal wall layer (ML) in the peg region cannot be clearly ascertained but may exist as a thin covering layer. The portion of the host wall (arrow) in advance of the peg stains more intensely than other regions. Note the multivesicular structures in the peg (V) and the appositions of material (A) on the host wall around the penetration site. (× 66,000) (Littlefield and Bracker, 1972.)

Figs. 29–31.   Sections 8, 5, and 1 respectively from a series of serial sections taken through a penetration peg developing from a haustorial mother cell (HMC) of *Uromyces phaseoli* var. *vignae.* The maximum development of the peg into the host cell wall (W) is shown in Fig. 30. Note the electron-opaque granules (arrow) always found in the peg at this stage of development. (All × 26,700) (From Heath and Heath, 1975.)

spite of the fact that many mention that these were specifically looked for. Ehrlich and Ehrlich (1971b) explain this in terms of the transient nature of the channels but to the authors of this book, it seems strange that of the hundreds of published and unpublished micrographs of haustoria which must exist, not one has been found to convincingly show such channels. Thus, to us, the evidence seems to indicate strongly the absence of any direct continuity between the cytoplasms of host and fungus. It should be pointed out, however, that there are precedents for such continuity between host and parasite in other situations (Dörr, 1969; Hoch, 1977). However, in these cases, both partners of the association belong to the same broad taxonomic group (i.e., vascular plants and fungi, respectively); protoplasmic continuity between such phylogenetically dissimilar organisms as the rusts and their hosts have not, to our knowledge, been documented.

2. Haustorium Development

The stages in the early development of the dikaryotic haustorium are rarely seen with the electron microscope, probably because of the ephemeral nature of certain stages of haustorial formation (Littlefield, 1972). The first recognizable event in haustorium formation is the development, from the dense, nonvacuolate, haustorial mother cell, of the penetration peg which breaches the host wall. Convincing early stages of formation of this peg (which definitely cannot be interpreted as oblique sections of the necks of well-developed haustoria) have been seen only for *Melampsora lini* (Littlefield and Bracker, 1972) and *Uromyces phaseoli* var. *vignae* (Heath and Heath, 1975). In both, the peg develops as a localized evagination of the haustorial mother cell plasmalemma and the *inner* layer of the haustorial mother cell wall (Fig. 28). This occurs in the center of the region apparently thickened by the addition of the middle electron-opaque wall layer but the contribution of this layer to the developing peg is unclear since it cannot be traced with any certainty beyond the point where it seems to merge with the evaginated inner layer (for further discussion on the fate and reality of this middle layer see Section II,A,1,a, this chapter). The outer wall layer, however, more clearly seems to play no part in peg development.

In *Uromyces phaseoli* var. *vignae,* all of the six penetration pegs seen in the process of bridging the host cell wall contain electron-opaque granules of about 30 nm in diameter (Figs. 29, 30, 31). The origin of these, or whether they are membrane bound, could not be determined but such structures do not resemble the multivesicular structures seen in penetration pegs of *Melampsora lini* (Fig. 28) (Littlefield and Bracker, 1972). Neither these vesicles nor the granules resemble the vesicles characteristic of hyphal apices (compare Figs. 28 and 30 with Fig. 13) which is perhaps surprising considering both structures are growing by "tip-growth." Conceivably, this difference

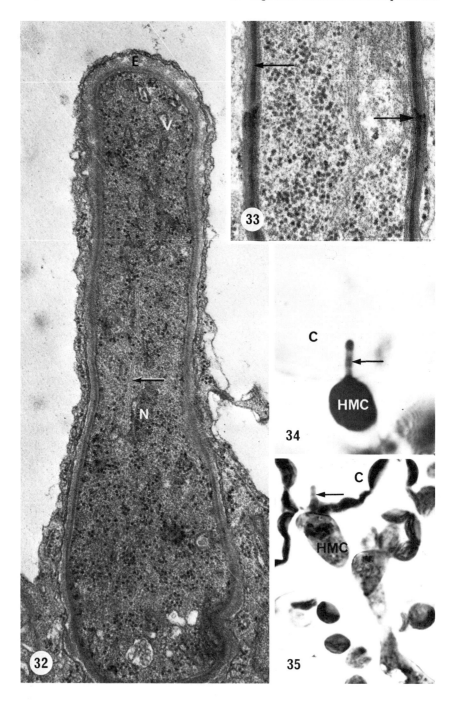

may reflect the fact that the penetration peg is also secreting cell-wall degrading enzymes since the lack of gross distortion of the fibrils of the host wall indicates a dissolution of the wall rather than mechanical penetration (see Section II,A,1,b, this chapter). However, claims of localized acid phosphatase activity in the penetration region (Bossányi and Oláh, 1974) are not supported by clear micrographs and, indeed, the vacuolate nature of the haustorial mother cells in these pictures suggests that the "penetration pegs" are oblique sections of a later stage of haustorium development. Ehrlich and Ehrlich (1963) have proposed that penetration is accomplished by a "blow-out" of the fungal protoplast resulting in a transient stage where the fungal wall is absent around the penetration peg. However, unequivocal ultrastructural evidence for this hypothesis is lacking.

The greatest amount of information on haustorium development beyond the penetration peg stage has so far come from work on *Uromyces phaseoli* var. *vignae* but even with this species the earliest stage examined has been a single example of a 4-μm-long haustorial neck which shows no sign of the development of a haustorial body (Fig. 32) (Heath and Heath, 1975). A few membrane-bound vesicles occur close to the tip of this neck but their morphology does not resemble that of either the apical vesicles of hyphal apices or the granules seen in the peg (compare Fig. 32 with Figs. 13 and 30). In fact, these vesicles may have little relationship to the extension of the neck since its length suggests that linear growth must be nearly completed. Ribosomes and longitudinally arranged microtubules are also present in the neck cytoplasm but mitochondria and nuclei are absent. The neck is com-

---

Fig. 32. An approximately median section of a developing haustorial neck (N) of *Uromyces phaseoli* var. *vignae*. Serial sections revealed no sign of a terminal haustorial body. Note the presence of microtubules (arrow) and vesicles (V) in the cytoplasm and the absence of mitochondria (these and the two nuclei are still in the haustorial mother cell). An extrahaustorial matrix (E) seemingly separates the fungal wall from the extrahaustorial membrane in all regions of the neck. No neckband can be detected. (× 41,000) (From Heath and Heath, 1975.)

Fig. 33. An early stage in neckband (large arrow) development in a *U. phaseoli* var. *vignae* haustorium in which the haustorial body is about 2.5 μm in diameter and the nuclei have not migrated from the haustorial mother cell. The neckband can be seen as an accumulation of electron-opaque granules bridging the fungal wall (small arrow) and the surrounding extrahaustorial matrix. (× 57,000) (From Heath and Heath, 1975.)

Fig. 34. Light micrograph of an acid fuchsin stained haustorial mother cell (HMC) of *Melampsora lini* which has formed a haustorial neck, but no body, in the host cell (C). Note the darkly-stained region about midway along the haustorial neck (arrow). (× 2,100) (From Littlefield, 1972. Reproduced by permission of the National Research Council of Canada from *Can. J. Bot.* 50, 1701–1703.)

Fig. 35. Phase-contrast, light micrograph of a haustorial mother cell (HMC) of *U. phaseoli* var. *typica* which has formed a haustorial neck in a host cell (C). Note the darker region about midway along the haustorial neck (arrow). This material has been prepared by standard techniques used for electron microscopy. (× 1,750) (From Mendgen, 1978.)

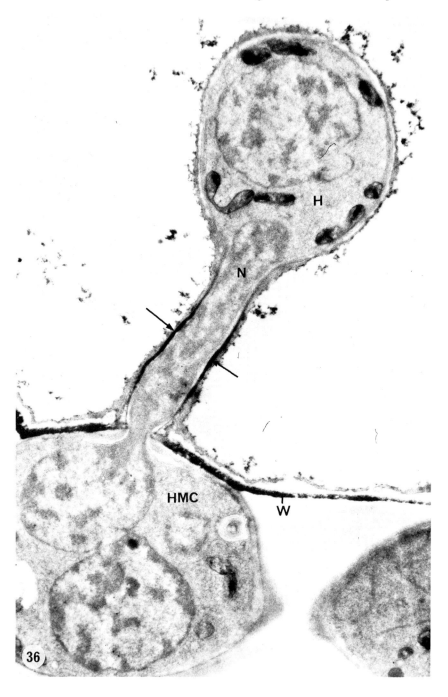

pletely surrounded by a discrete fungal wall of the same thickness (18–20 nm) as neck walls of mature haustoria, and this wall is separated from the host plasmalemma along its entire length by an extrahaustorial matrix of moderate electron opacity. No neckband can be detected, nor is there any morphological indication of any sort of "tight junction" between the fungal and host plasmalemmas anywhere along the neck.

Stages in the formation of the haustorial body of *Uromyces phaseoli* var. *vignae* are seen more frequently than stages in neck development (Heath and Heath, 1975). The initial swelling of the body apparently coincides with the migration of mitochondria from the haustorial mother cell. In the neck, these organelles show a close morphological association with preformed cytoplasmic microtubules, suggesting the latter may act as "tracks" for organelle migration as has been postulated for other organisms (see Heath and Heath, 1971, 1975). The neckband becomes detectable when the haustorial body is 2.5–3.0 μm in diameter (Fig. 33); its development precedes the migration [again associated with cytoplasmic microtubules (Fig. 24), Heath and Heath, 1971] of the two nuclei and most of the remaining haustorial mother cell cytoplasm, into the haustorium (Heath and Heath, 1975).

Until the migration of the nuclei, the haustorial body of *U. phaseoli* var. *vignae* (Heath and Heath, 1971) and *Puccinia graminis* f. sp. *tritici* (Harder *et al.*, 1978) is covered by a discrete fungal wall similar in thickness to, and continuous with, that covering the neck region. For the former fungus and *U. phaseoli* var. *typica* (Chapter 6, Fig. 5), the larger body seen after nuclear migration has a much less discrete and often barely visible wall suggesting that this phase of enlargement occurs rapidly and is accompanied by stretching, rather than synthesis, of the fungal wall. Wall synthesis apparently follows however, since this portion of the wall soon becomes discrete and the same thickness as that around the haustorial neck (Fig. 37). In *U. phaseoli* var. *vignae*, the body wall eventually seems to contain two layers (Fig. 38) (see discussion in Section II,A,1,d, this chapter) but these layers appear to merge into one at the junction with the haustorium neck. It is interesting that during no stage of haustorial body formation in this fungus is the host plasmalemma tightly appressed to the fungal wall; thus an extrahaustorial "matrix" is present at all times although this always has a more

Fig. 36. The first-formed haustorium (H) of *Uromyces phaseoli* var. *typica* showing a nucleus (N) migrating into the haustorium from the haustorial mother cell (HMC). Cytochemical localization of peroxidase activity, indicated by electron-opaque deposits, reveals such activity on the host wall (W) (which is also present in uninfected cells) and in a broad band (arrows) around the haustorial neck. ($\times$ 13,600) [From Mendgen, 1975. Reproduced with permission from *Physiol. Plant Pathol.* **6**, 275–282. Copyright Academic Press, Inc. (London), Ltd.]

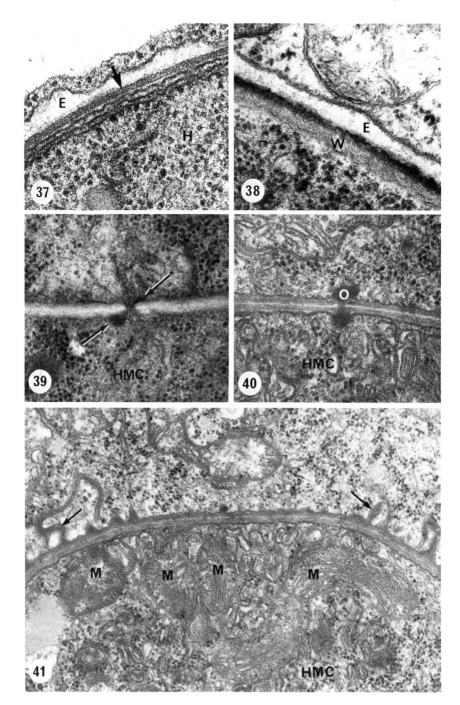

electron-lucent appearance than the corresponding region along the haustorium neck (see Section II,A,1,d, this chapter for discussion of the possible artifactual nature of this electron-lucent matrix).

In *Uromyces phaseoli* var. *vignae,* haustorium formation is characterized by intriguing changes associated with the septum delimiting the haustorial mother cell (Heath and Heath, 1975). The septal pore is originally open and there is no pore apparatus (Fig. 39). The pore begins to become plugged with electron-opaque material before the development of the penetration peg (Figs. 39 and 40) and during this process, tubular protrusions develop by invagination of the plasmalemma on the hyphal side of the septum (Fig. 41). These grow to about 400 nm in length by the time the pore has become completely blocked at which stage mitochondria accumulate on the haustorial mother cell side of the septum and intermingle with masses of membranes apparently formed by elaborations of the adjacent plasmalemma (Fig. 41). These membranes disappear and the mitochondria become dispersed by the time the penetration peg begins to enter the host wall but the plasmalemma protrusions on the hyphal side of the septum continue to elongate to their maximum length of about 900 nm. Their maximum development is accompanied by a marked association with numerous mitochondria and coincides with the penetration of the host wall (Fig. 42). By the time the haustorial neck has reached its more or less full length, however, the mitochondria have dispersed and the plasmalemma protrusions have shrunk to less than 400 nm.

Fig. 37.   A young, first-formed haustorium (H) of *Uromyces phaseoli* var. *vignae.* Note the discrete nature of the haustorial wall (arrow) and the surrounding extrahaustorial matrix (E). (× 79,400) (M. C. Heath, unpublished.)

Fig. 38.   An older haustorium of *U. phaseoli* var. *vignae* in an established infection. The fungal wall (W) now appears to consist of two layers (compare with Fig. 37) both of which are reasonably distinct from the surrounding extrahaustorial matrix (E). (× 84,200) [From Heath and Heath, 1971. Reproduced with permission from *Physiol. Plant Pathol.* **1**, 277–287. Copyright Academic Press, Inc. (London), Ltd.]

Fig. 39.   Median section through the pore of a haustorial mother cell (HMC) septum at the earliest stage found prior to haustorium formation. Note the continuity of cytoplasm through the pore and the electron-opaque material (arrows) associated with adjacent portions of the septum. (× 58,000) (From Heath and Heath, 1975.) For a diagrammatic representation of the changes in haustorial mother cell septum of *Uromyces phaseoli* var. *vignae* shown in Figs. 39–44, see Fig. 19.

Fig. 40.   The pore of the haustorial mother cell (HMC) septum (*Uromyces phaseoli* var. *vignae*) shown in Fig. 41. Note the occlusion of the pore with electron-opaque material (O). The haustorium has not yet started to develop. (× 58,000) (From Heath and Heath, 1975.)

Fig. 41.   Another section of the septum shown in Fig. 40. Note the accumulation of mitochondria (M) and the abundance of membrane profiles on the haustorial mother cell (HMC) side of the septum. Tubular protrusions of the plasmalemma (arrows), filled with electron-opaque material, can be seen on the hyphal side of the septum. (× 42,000) (M. C. Heath and I. B. Heath, unpublished.)

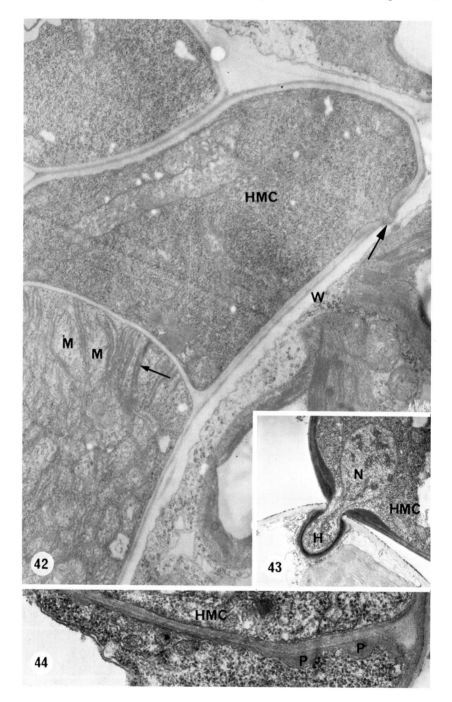

Little or no sign of these protrusions can be seen by the time both nuclei have entered the haustorium body (Fig. 44). During nuclear migration, electron-lucent areas appear in the septal pore plug and differentiated regions of cytoplasm develop on either side of the septum. Microbodies aggregate around the periphery (probably by migration since no sign of synthesis was observed) and the septal pore apparatus now resembles that typical of other parts of the intercellular mycelium. These changes associated with the haustorial mother cell septum, and the corresponding development of the haustorium, are shown in diagrammatic form in Fig. 19.

In terms of morphology and changes in wall structure, haustorium forma-tion from the haustorial mother cell of *Uromyces phaseoli* var. *vignae* is analogous to substomatal vesicle formation from the appressorium (see Chapter 3). However, in contrast to vacuolation in the appressorium of the same fungus (see Chapter 3, Section III,D) dense-body vesicles are absent in the haustorial mother cell and can therefore have no role in the vacuolation which accompanies cytoplasmic migration into the haustorium (M. C. Heath, unpublished).

The ultrastructural features of D-haustorium formation and the accompa-nying changes associated with the haustorial mother cell in *Uromyces phaseoli* var. *vignae*, suggest that the D-haustorium is not a mere "extension" of the intercellular mycelium but is a highly specialized structure whose for-mation requires certain complex physiological processes. By analogy to other situations where mitochondria are associated with plasmalemma elaborations (e.g., transfer cells of higher plants), the sequential development of extra membrane on either side of the haustorial mother cell septum suggests that additional membrane area is needed at this time to facilitate energy-requiring transport of materials across the septum, first into the haustorial mother cell and then out of it (Heath and Heath, 1975). If such transport is

---

Fig. 42.   A haustorial mother cell (HMC) of *Uromyces phaseoli* var. *vignae* in the process of forming a penetration peg (large arrow). Serial sections revealed that the peg had just breached the host wall (W). Note the mitochondria (M) associated with the elongated plasmalemma pro-trusions (small arrow) on the hyphal side of the haustorial mother cell septum. The association of membranes and mitochondria on the other side of the septum (see Fig. 41) has now disappeared. (× 20,800) (From Heath and Heath, 1975.)

Fig. 43.   A haustorial mother cell (HMC) from which the last of the two nuclei (N) is ap-parently in the process of migrating into the obliquely sectioned haustorial neck (H). At this stage of haustorium development, the haustorial mother cell septum begins to develop the septal pore apparatus (see Fig. 2) typical of intracellular hyphae, and the plasmalemma protrusions have virtually disappeared. (× 10,600) (From Heath and Heath, 1975.)

Fig. 44.   The edge of a haustorial mother cell (HMC) septum at a slightly earlier stage in haustorium development than that shown in Fig. 43. The remains of the plasmalemma protru-sions (P) can be recognized only as small mounds of electron-opaque material. (× 37,600) (From Heath and Heath, 1975.)

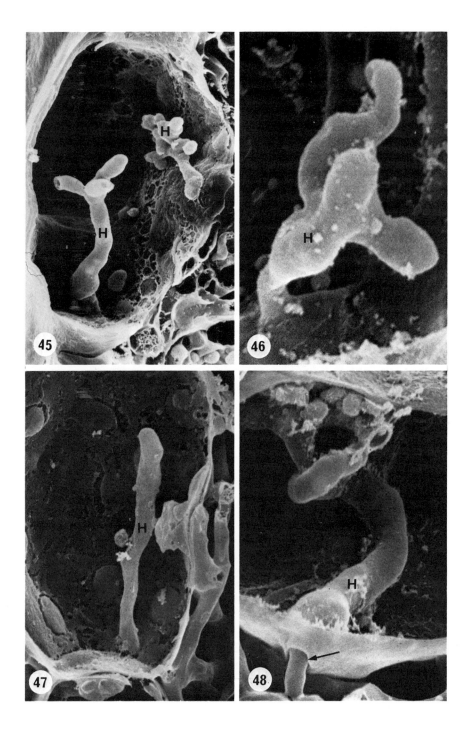

Figs. 45–52. The mature, monokaryotic (M) haustorium.

occurring, the observed plugging of the septal pore during this process would be a necessity to prevent redistribution of whatever was being transported; however, just what the haustorial mother cell requires, or needs to have removed, only at the time of host penetration, is difficult to surmise.

It remains to be seen whether such changes in the haustorial mother cell, or the above-described stages of haustorium formation, occur in all other rusts. So far, the available evidence suggests that this is the case since plasmalemma protrusions on the hyphal side of the haustorial mother cell septum at the time of penetration have been observed in *Uromyces phaseoli* var. *typica* at the time of host penetration (M. C. Heath, unpublished), in *Puccinia hordei* just before penetration (Reynolds, 1975), and can be discerned in the micrograph illustrating the penetration peg of *Melampsora lini* (Fig. 2 of Littlefield and Bracker, 1972). Similarly, the absence of a neck-band from the young haustorium has been shown for *P. graminis* f. sp. *tritici* (Fig. 58) (Harder *et al.*, 1978), *U. phaseoli* var. *typica* (Mendgen, 1973a) and *M. lini* (Coffey, personal communication). These latter observations, however, seem to conflict with the fact that "neckbands" have been seen by light microscopy around bodyless necks of *M. lini* after staining (Fig. 34) (Littlefield, 1972) and of *U. phaseoli* var. *typica* under phase contrast (Fig. 35) (Mendgen, 1978). Since, at least for *U. phaseoli* var. *typica,* the material was prepared for light microscopy in the same manner as material prepared for electron microscopy, this conflict cannot be explained on the basis of different treatment of the tissue. Thus it seems likely that the neckband seen with the light microscope is not the same structure as that visualized by electron microscopy. Possibly the role of the latter is solely to restrict apoplastic transport along the neck since its absence correlates well with the absence of any restriction of such transport in the young haustorium (Heath, 1976b). The neckband seen with the light microscope may be the visual sign of a completely different phenomenon which may be related to the broad band of peroxidase activity shown to encircle the necks of *Uromyces phaseoli* var. *typica* haustoria (see Section II,A,1,c, this chapter) before (Fig. 36) and after the much narrower neckband develops (Mendgen, 1975). One obvious conclusion from these observations, however, is that our knowledge of the interactions between host and fungus in the haustorial neck region is far from complete.

Figs. 45–48. Scanning electron micrographs of mature M-haustoria (H) of *Puccinia recondita* (Fig. 45), *P. malvacearum* (Figs. 46 and 48) and *Melampsora lini* (Fig. 47). Note their hyphalike morphology and the common absence of differentiation into body and neck regions (compare with Fig. 21). In Fig. 48, the way in which the host cell wall protrudes outwards (arrow) suggests that the haustorium is exiting from the cell. (Fig. 45, × 1,900; Fig. 46, × 12,000; Fig. 47, × 2,400; Fig. 48, × 5,400) [Fixed in glutaraldehyde. Figs. 45 and 47 from Gold *et al.,* 1979. Reproduced by permission of the National Research Council of Canada from *Can. J. Bot.* (in press); Figs. 46 and 48, R. E. Gold and L. J. Littlefield, unpublished.]

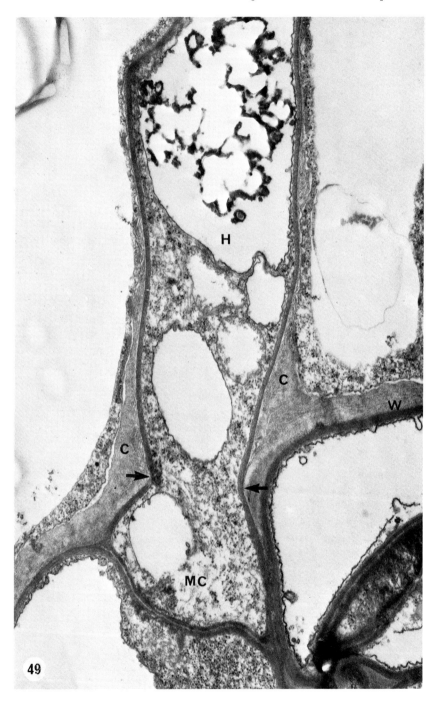

B. Monokaryotic Haustoria

The first ultrastructural studies of pycnial and aecial rust infections (Boyer and Isaac, 1964; Longo and Naldini, 1970; Moore and McAlear, 1961; Orcival, 1969) revealed intracellular structures which, in cross section, seemed to closely resemble the haustoria seen in uredial and telial stages. However, more recent investigations suggest that monokaryotic (M) (i.e., found in basidiospore-derived infections, see footnotes to Section II, this chapter) and dikaryotic (D) haustoria are not ultrastructurally identical. Several light microscopists have previously reported that M-haustoria are more filamentous and irregular in shape than D-haustoria (Allen, 1932b; Colley, 1918; Dodge, 1922; Pady, 1935a,b) and electron microscopy has supplemented these studies by showing more clearly that, for all species examined, M-haustoria lack any clear differentiation of a neck region (Figs. 45–48) (*Cronartium ribicola*, Robb *et al.*, 1975b; *Gymnosporangium haraeanum*, Kohno *et al.*, 1976; *Kuehneola japonica*, Kohno *et al.*, 1977a; *Melampsora lini*, Gold and Littlefield, unpublished; *M. pinitorqua*, M. D. Coffey, personal communication; *Peridermium pini*, Walles, 1974; *Puccinia coronata* f. sp. *avenae*, Harder, 1978; *P. recondita*, Gold *et al.*, 1979; *P. sorghi*, Rijkenberg and Müller, 1971; Rijkenberg and Truter, 1973a; *Uromyces phaseoli* var. *vignae*, M. C. Heath, unpublished) except where apparently constrained by a "collar" (Harder, 1978; Kohno *et al.*, 1977a; Robb *et al.*, 1975b; Walles, 1974; ultrastructural details of these collars are given in Section II,D,2, of this chapter). In all these studies, the fungus also shows relatively little constriction as it passes through the host wall (Fig. 49), although, as in walls invaded by the much narrower penetration pegs of D-haustoria (see Section II,A,1,b, this chapter), the lack of fibril distortion, or the presence of signs of degradation suggests penetration to be enzymatic rather than purely mechanical (Harder, 1978; Rijkenberg and Truter, 1973a; Walles, 1974).

Monokaryotic and dikaryotic haustoria also seem to be ultrastructurally distinguishable by the fact that the M-haustorial wall is of the same thickness as that of the penetration peg and mother cell, and no localized thickening of the latter, or differential development of wall layers (compare with D-haustoria, Section II,A,1,d, this chapter), occurs at the junction of the

---

Fig. 49. Transmission electron micrograph of an M-haustorium (H) of *Uromyces phaseoli* var. *vignae*. Note the small size of the mother cell (MC) and the only slight constriction (arrows) as the fungus passes through the host cell wall (W) (compare with Figs. 22–24). The fungal wall is of the same thickness throughout the penetration region and no electron-opaque neckband can be seen anywhere along the haustorium (compare with Figs. 22–24). A fibrillar, Type II collar (C) (for diagram see Fig. 20), which seems continuous with the host wall, surrounds the fungus as it enters the cell. An extrahaustorial matrix seems to be absent from this end of the haustorium but can be found in its more distal regions (see Fig. 52). (× 23,500) (D. M. Tighe and M. C. Heath, unpublished.)

mother cell with the haustorium (Gold *et al.*, 1979; Harder, 1978; Kohno *et al.*, 1977a; Rijkenberg and Truter, 1973a; Robb *et al.*, 1975b; Walles, 1974). In *Uromyces phaseoli* var. *vignae*, the walls of both mother cell and haustorium respond similarly to periodic acid–chromic acid–phosphotung-stic acid (PACP) staining (M. C. Heath, unpublished) unlike the situation for D-haustoria of *Melampsora lini* (Fig. 26) (Littlefield and Bracker, 1972). While the mother cells of M-haustoria resemble those of D-haustoria in their separation from the rest of the mycelium by a perforate septum associated with a typical septal pore apparatus (*U. phaseoli* var. *vignae*, D. M. Tighe and M. C. Heath, unpublished), light microscopy (e.g., R. F. Allen, 1932b, 1935; Colley, 1918) suggests that the monokaryotic (M)-mother cell may not be terminal. The suggestion that M-haustoria of *Puccinia coronata* f. sp. *avenae* lack any attachment of the M-mother cell to the host wall (Harder, 1978) does not seem to be a universal phenomenon since material resembling that thought to have an adhesive role in dikaryotic infections occurs at the penetration site of M-haustoria of *P. sorghi* (Rijkenberg and Truter, 1973a) and *U. phaseoli* var. *vignae* (D. M. Tighe and M. C. Heath, unpublished).

Another significant difference between M- and D-haustoria seems to be the absence of a densely staining neckband (Fig. 49) in conventionally prepared M-haustoria of *Kuehneola japonica* (Kohno *et al.*, 1977a), *Melampsora pinitorqua* (M. C. Coffey, personal communication), *Peridermium pini* (Walles, 1974), *Puccinia coronata* f. sp. *avenae* (Harder, 1978), *P. sorghi* (Rijkenberg and Truter, 1973a), and *Uromyces phaseoli* var. *vignae* (M. C. Heath, unpublished); in the latter species no neckband is revealed by periodate treatment (compare with D-haustoria, Section II,A,1,c, this chapter). Only for M-haustoria of tissue-culture grown *Cronartium ribicola* have electron-opaque, neckbandlike structures been reported (Robb *et al.*, 1975b) and since these appear continuous with extensive electron-opaque deposits in the host cytoplasm, and are only seen when the fungus is necrotic, it seems possible that they are artifacts associated with fungal necrosis; certainly they do not seem to represent the same structure seen in the dikaryotic haustorial apparatus.

Whether a *functional* neckband (i.e., some barrier to apoplastic transport) exists around M-haustoria is still unknown but an extensive search for any morphological "tight junction" between the extrahaustorial membrane and the fungus in pycnial infections of *Uromyces phaseoli* var. *vignae* has con-sistently failed to reveal such a structure (M. C. Heath, unpublished).

Both light (R. F. Allen, 1935; Pady, 1935a) and electron microscopy have revealed that many M-haustoria are septate. Haustoria of *Cronartium ribicola* (Robb *et al.*, 1975b), *Kuehneola japonica* (Kohno *et al.*, 1977a), *Melampsora pinitorqua* (M. D. Coffey, personal communication), and *Peridermium pini* (Walles, 1974) always seem to have a septum in the penetration region, either within the limits of the host cell wall or just inside the cell lumen (Fig. 51). In *P. pini* infections some sections show the septum to be perforate but plugged with electron-opaque material (Walles, 1974), but in *M. pinitorqua* both plugged and unplugged pores have been seen and the latter have a septal pore apparatus similar to that found in vegetative mycelium (M. D. Coffey, personal communication). No similar information has been reported for the other species, probably due to a lack of median sections rather than an absence of a pore. Reports of septa in other species (*Gymnosporangium haraeanum*, Kohno *et al.*, 1976; *Puccinia coronata* var. *avenae*, Harder, 1978; *P. sorghi*, Rijkenberg and Truter, 1973a) suggest that they normally occur in a more distal portion of the haustorium relative to the penetration site (Fig. 50) and are not a consistent feature.

Like D-haustoria, M-haustoria invaginate the host plasmalemma (Fig. 49) (Harder, 1978; Kohno *et al.*, 1977a; Orcival, 1969; Rijkenberg and Truter, 1973a; Walles, 1974) and the fungal wall usually seems to be somewhat separated from the surrounding extrahaustorial membrane (Fig. 49). In *Cronartium ribicola* (Robb *et al.*, 1975b), *Peridermium pini* (Walles, 1974), and *Uromyces phaseoli* var. *vignae* (M. C. Heath, unpublished) such separation is usually more pronounced at the distal portions of the haustorium as well as close to the site of penetration. However, the material between the fungal wall and the extrahaustorial membrane commonly looks different in these two regions which has led most authors to consider the material closest to the host wall as a host-derived collar and the rest as some type of extrahaustorial matrix of uncertain origin and, possibly, different composition. However, Rijkenberg and Truter (1973a) do not make such a distinction for *Puccinia sorghi* infections and consider the whole haustorium to be covered by a continuous layer of either collar or matrix. Since their micrographs do not resemble those of *Peridermium pini* infections where some haustoria are very clearly encased in material continuous with the collar (and where an extrahaustorial matrix seems also to be present) (Walles, 1974), it seems likely that most of the material surrounding *P. sorghi* haustoria is equivalent to that interpreted as matrix in other rust infections, and that it blends imperceptibly with the collar at the region of penetration. However, both collar and matrix have been suggested to contain polysaccharides (see Host Responses, Section II,C,2, this chapter) and it is possible that there is a greater similarity in composition between these structures in *P. sorghi* infections than found in other rust–host interactions.

In most of the studies mentioned above, the extrahaustorial matrix is relatively electron-opaque (Figs. 49 and 52) although in some areas where it is particularly wide, it often appears more heterogeneous with both translucent and opaque components (Figs. 52 and 67); however, more uniformly translucent matrices have been reported for *Gymnosporangium haraeanum*

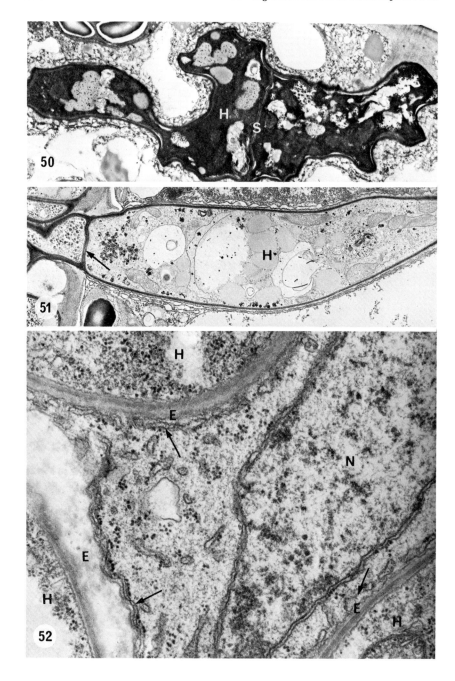

(Kohno *et al.*, 1976), *Puccinia coronata* f. sp. *avenae* (Harder, 1978), and *P. urtici-caricis* (Orcival, 1969). While, as described earlier, the width of the matrix in some interactions seems to depend on the distance from the penetration site, in *Kuehneola japonica* it is claimed that no matrix can be seen before 20 days after inoculation (Kohno *et al.*, 1977a). The question of whether the matrix is a fixation artifact arises again (see Section II,A,1,d, this chapter for a discussion of this in relation to the D-haustorium) and, again, there is no unequivocal answer, although the frequent presence of stainable material suggests that there is something real separating the extrahaustorial membrane from the fungal wall, even if the separation has been accentuated during fixation. The nature of this staining material is essentially unknown although in *Uromyces phaseoli* var. *vignae* it stains strongly after PACP treatment (Fig. 54) which may indicate the presence of glycoprotein or carbohydrate (Rambourg, 1967; Roland, 1969; for a discussion of the specificity of this stain see Section II,A,1,d, this chapter). Whether this material is of host or fungal origin is also uncertain. Boyer and Isaac (1964) suggest it to be fungal in *Cronartium ribicola* infections due to its resemblance to the material which surrounds intercellular hyphae; however, Robb *et al.* (1975b) claim that these layers are *not* continuous through the penetration region. The slight evidence that this matrix material may be of host origin is discussed under Host Responses (Section II,C,2, this chapter).

A rather different type of extrahaustorial matrix to those described above has been reported for some haustoria of *Kuehneola japonica* (Kohno *et al.*, 1977a). This type is wide, primarily electron-lucent, and contains electron-opaque, vesicular profiles, and thus closely resembles callose-containing collars and encasements of both M- and D-haustoria (see Figs. 80 and 81). However, the authors suggest that it differs from these in that it is deposited directly into the extrahaustorial matrix region rather than being separated

---

Fig. 50. An M-haustorium (H) of *Puccinia coronata* f. sp. *avenae*. Note its irregular shape and dense contents which *may* be an artifact of specimen preparation (see text). A septum (S) is present in the haustorium but this is not a consistent feature. (× 9,500) (From Harder, 1978. Reproduced by permission of the National Research Council of Canada from *Can. J. Bot.* **56,** 214-224.)

Fig. 51. An M-haustorium (H) of *Melampsora pinitorqua* showing a septum (arrow) at the penetration site; its presence here seems characteristic for this species. (× 8,100) (courtesy of M. D. Coffey, unpublished.)

Fig. 52. Three M-haustorium (H) profiles of *Uromyces phaseoli* var. *vignae* in the same host cell (it is unknown whether they are part of one haustorium). Note the differences in appearance and width of the extrahaustorial matrices (E). The flocculent material in the bottom, left matrix may be equivalent to that which stains after PACP treatment (see Fig. 54). All three haustorium profiles are associated with endoplasmic reticulum (arrow) and a lobe of the host nucleus (N). (× 41,200) (M. C. Heath, unpublished.)

from the fungal wall by two host-derived membranes. The validity of this interpretation is questioned however, and is discussed further under Host Responses (Section II,C,2, this chapter).

All of the studies described above show that, like D-haustoria, M-haustoria have the same general cytoplasmic components as the other parts of the vegetative mycelium. Harder (1978) reports the M-haustoria of *Puccinia coronata* f. sp. *avenae* are generally more densely staining and more irregular in outline (Fig. 50) than D-haustoria of the same species but such a feature is not apparent from other studies and may indicate a difference in response to preparative procedure rather than any absolute difference in cytoplasm or shape. Harder (1978) also reports that glycogen occurs in M-, but not D-, haustoria of this rust but no similar phenomenon occurs in *Uromyces phaseoli* var. *vignae* (M. C. Heath, unpublished). In the latter fungus, one of the few cytoplasmic differences between the mature M- and D-haustoria seems to be the more frequent presence of microtubules in the former and it is conceivable that this relates to organelle transport within the longer and much more filamentous structure. Unfortunately, comparisons of M- and D-haustoria of the same rust species are too few yet to tell whether any of the above-mentioned observations are general phenomena or just limited to the species examined. One feature which may be fairly widespread, however, is the variability in the number of nuclei present since one, *or more,* have been reported for M-haustoria of several species (Kohno *et al.,* 1976, 1977; Rijkenberg and Truter, 1973a; Robb *et al.,* 1975b).

A striking gap in our ultrastructural knowledge of M-haustoria is their mode of formation. At present there is no information on whether the mother cell shows any special differentiation during development. It is also unknown whether the cytoplasm migrates from the mother cell to the haustorium [although R. F. Allen's (1932b) light microscope study suggests the nucleus, at least, does so] or whether the large size of the haustorium means that additional cytoplasm is synthesized during haustorium growth. Equally, it is unknown whether the M-haustorium grows by tip growth and whether its growth is ultrastructurally more similar to that of hyphal apices or D-haustoria. It is hoped that more detailed ultrastructural studies of the early stages of haustorium development will provide answers to these questions.

Taken as a whole, the available ultrastructural studies suggest that M-haustoria show little of the structural specialization characteristic of the D-haustorial apparatus. In addition, in *Gymnosporangium haraeanum* (Kohno *et al.,* 1976) and *Kuehneola japonica* (Kohno *et al.,* 1977a), there seems to be little difference between M-haustoria and the intracellular structures formed during invasion of epidermal cells by basidiospore germ tubes. Thus for these species there appears to be no marked ultrastructural distinction between those intracellular structures destined to leave the cell, and those which may show delimited growth within it. This raises the point, however, of just how delimited is the M-haustorium. This is often a difficult question to answer from observations of thin sections although Coffey (personal communication), Harder (1978), Robb *et al.* (1975b), and Walles (1974) all claim, or seem to regard, the intracellular structures observed in their studies as restricted to a single host cell. In contrast, scanning electron microscopy has shown M-haustoria of *Puccinia malvacearum* (Fig. 48), and *P. recondita* (Gold *et al.*, 1979) apparently leaving the host cell. However, neither transmission, nor scanning electron microscopy is the best tool for examining the range of morphologies found among M-haustoria. The best information comes from light microscopy and the few available studies show that while the M-haustoria of some species are characteristically restricted to one host cell (e.g., *Cronartium ribicola* in parenchyma cells, Colley, 1918; *Puccinia coronata*, Allen, 1932b; *Uromyces phaseoli* var. *vignae*, M. C. Heath, unpublished), those of other species may ramify into other cells or intercellular spaces (e.g., *Cronartium ribicola* in the phloem, Colley, 1918; *Gymnoconia interstitialis*, Pady, 1935b; *Melampsora lini*, Gold and Littlefield, 1979; *P. malvacearum*, R. F. Allen, 1935; *Ravenelia* spp., Rajendren, 1972). Thus it appears that some of these intracellular structures conform to Bushnell's (1972) definition of a haustorium but others do not. For this reason, and because of their ultrastructural differences from D-haustoria, several authors (Kohno *et al.*, 1976; Rijkenberg and Truter, 1973a) have preferred to consider all these intracellular structures as "intracellular hyphae," regardless of morphology. We have chosen tentatively to use the term "haustorium" in this book in order to point out that both monokaryotic and dikaryotic intracellular structures represent analogous portions of the fungus in terms of the close association with the host protoplast. Which of these terms is the best ultimately rests on how similar these structures are in function (see Concluding Remarks, Section II,D) and whether either type differs from intracellular hyphae of other parasitic fungi. Until this information is available, perhaps a descriptive approach should be adopted and both "intracellular hypha" and "haustorium" used depending on the morphology of the structures examined; "intracellular structure" could be used as a noncommittal term if sufficient morphological information is lacking.

C. Host Responses

1. Cytoplasmic Changes

The host response to initial infection by rust fungi has received little ultrastructural study, but the few observations of young, intercellular, infection hyphae derived from urediospore inoculations suggest that the surrounding host cells differ little, if at all, from those in uninfected leaves (Heath and Heath, 1971; Mendgen, 1975; Onoe *et al.*, 1976). Thus, in these host species at least, responses by susceptible plants seem to occur only after formation of haustoria (however, in well established infections, both invaded and non-invaded cells may be affected). There has been only one study of the direct penetration of epidermal cells by basidiospore-derived penetration pegs (*Kuehneola japonica*, Kohno *et al.*, 1977a) and such infection results in the complete disruption of the invaded cell; whether this is a general phenomenon is as yet unknown.

Most of the ultrastructural investigations of host responses to rust infection have focused on the interface between the M- or D-haustorium and the host cell and particularly on the invaginated portion of the host plasmalemma (the extrahaustorial membrane). Interest in this membrane is not surprising when one considers it theoretically controls not only the host-derived substances available to the haustorium but also the entry into the host protoplast of fungal materials.

The suggestion that the host plasmalemma is dissolved at the time of host cell penetration by the young D-haustorium (Bossányi and Oláh, 1972) is not supported by clear ultrastructural evidence (in fact the relevant micrograph seems to show a hole in the section at the crucial point), and features of the relevant stages in M-haustorium formation have not yet been reported. However, even around mature haustoria, unequivocal continuity between the extrahaustorial membrane and the host plasmalemma, or the intact nature of the extrahaustorial membrane, are often difficult to illustrate in either type of haustorial apparatus. As Bracker and Littlefield (1973) point out, this does not necessarily mean that either membrane is discontinuous since any membrane will appear blurred or diffuse if sectioned obliquely or tangentially. Thus the fact that continuity has been demonstrated in some cases (e.g., Figs. 23 and 49), and the absence of any convincing evidence to

---

Figs. 53–57 (continued on page 169). The extrahaustorial membrane.

Fig. 53. A D-haustorium (H) of *Melampsora lini* in a section treated with periodic acid–chromic acid–phosphotungstic acid (PACP). The fungal plasmalemma (large solid arrow) and the portion of the host plasmalemma (small solid arrow) against the host cell wall (W) stain more intensely than the invaginated portion of host plasmalemma (open arrow) (the extra-haustorial membrane) and other membranes of the cell. (× 70,000) (From Littlefield and Bracker, 1972.)

Fig. 54. A PACP-treated section showing two profiles of M-haustoria (H_1 and H_2) of *Uromyces phaseoli* var. *vignae* which may, or may not, be part of the same haustorium. The extrahaustorial membrane (arrow), and the material within the extrahaustorial matrix (E) of H_1 stain intensely while those of H_2 do not. The stained material in the matrix of H_1 may be equivalent to the flocculent, electron-opaque material seen in some matrices after conventional staining (see Figs. 52 and 67). (× 62,200) (M. C. Heath, unpublished.)

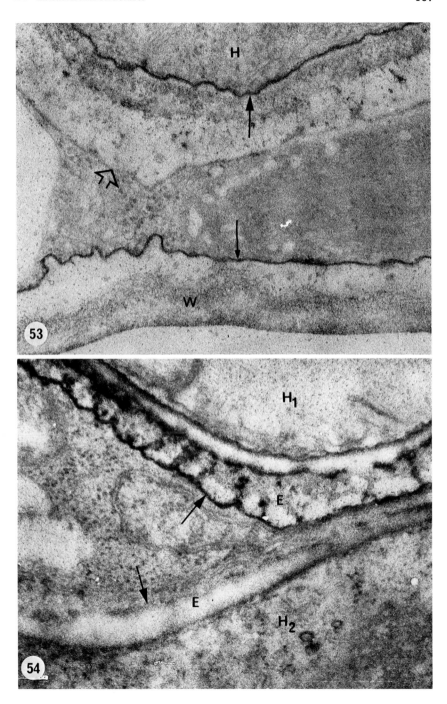

the contrary, suggests that the host plasmalemma is connected to a continuous extrahaustorial membrane in all invaded cells which are not obviously disorganized. There is, however, some convincing ultrastructural evidence for some rust infections that the invaginated part of the host plasmalemma differs in some way from the noninvaginated portion. In uredial infections of both *Hemileia vastatrix* (Rijo and Sargent, 1974) and *Melampsora lini* (Figs. 55 and 56) (Littlefield and Bracker, 1972), thin sectioning and freeze-etching, respectively, have revealed unusual furrows and undulations in the extrahaustorial membrane while thin sections of uredial infections of *Puccinia graminis* f. sp. *tritici* (Ehrlich and Ehrlich, 1963; D. E. Harder, personal communication) and *Uromyces phaseoli* var. *typica* (M. C. Heath, unpublished), show this membrane to be thicker than the rest of the host plasmalemma. In contrast, a thinner extrahaustorial membrane has been reported for M-haustoria of *P. coronata* f. sp. *avenae* in cases where the extrahaustorial matrix seemed poorly developed (Harder, 1978). However, none of these phenomena seem ubiquitous since differences in membrane thickness have not been detected in *Melampsora lini* (Littlefield and Bracker, 1970, 1972) or *Puccinia sorghi* (Van Dyke and Hooker, 1969) infections, and neither this feature, nor membrane furrows, have been even mentioned in other studies. Nevertheless, such lack of differentiation after conventional preparation does not necessarily indicate that such differentiation does not exist, since freeze-etching, or treatment with periodic acid–chromic acid–phosphotungstic acid (PACP) stain, respectively, reveals that the membrane around D-haustoria of *M. lini* bears fewer granules (Figs. 55–57) and stains less intensely (Fig. 53) than the rest of the host plasmalemma (Littlefield and Bracker, 1972). A similar lack of PACP staining of the extrahaustorial membrane occurs around M-haustoria of *Uromyces phaseoli* var. *vignae* (Fig. 54) and it is interesting that in this and the *M. lini* study, the decrease in stainability of the host membrane occurs at a point somewhat distant from the junction of the extrahaustorial membrane and the rest of the host plasmalemma. Around the D-haustoria of *M. lini,* the change more or less coincides with the position of the neckband (Littlefield and Bracker, 1972), but it is more variable, and does not coincide with any

---

Fig. 55. Freeze-etched preparation of a D-haustorium (H) and haustorial mother cell (HMC) of *Melampsora lini.* Note the undulations and depressions (arrows) in the extrahaustorial membrane. ($\times$ 17,400) (From Littlefield and Bracker, 1972.)

Fig. 56. Freeze-etched, face view of the extrahaustorial membrane surrounding an *M. lini* D-haustorium. Note the lack of granularity (compare with Fig. 57) and the cleftlike depressions in the membrane. ($\times$ 70,000) (From Littlefield and Bracker, 1972.)

Fig. 57. Freeze-etched, face view of the noninvaginated host plasmalemma. Note the increased granular texture compared with Fig. 56. ($\times$ 70,000) (From Littlefield and Bracker, 1972.)

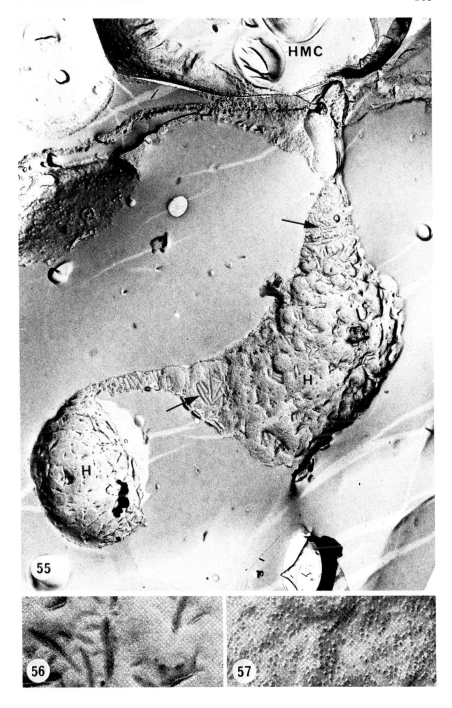

structural feature in M-haustoria of *U. phaseoli* var. *vignae* which lack the neckband (M. C. Heath, unpublished).

Littlefield and Bracker (1972) conclude from their *M. lini* study that either the change in nature of the invaginated plasmalemma occurs after its formation, or that the new portion of the plasmalemma differs from the rest from the outset; either alternative, they suggest, represents a response to the pathogen. It should be pointed out, however, that it is unknown whether the change in PACP response, or the lack of granules, is typical of *any* newly-formed portion of the plasmalemma induced to form in the absence of cell growth or wall synthesis. Without this information, the role of the fungus in this phenomenon can only be a matter of speculation. In this context, it is of interest to consider the nature of the substances stained by the PACP treatment. In one of the first published descriptions of this technique, Rambourg (1967) suggests that it probably stains glycoproteins. In support of this, there does seem to be some (but not complete) similarity in the images obtained by PACP staining and those produced by other techniques supposed to localize polysaccharides (Roland, 1969), even though the affinity of animal plasmalemmas for phosphotungstic acid (without the chromic acid) can be lost by protease treatment (Benedetti and Bertolini, 1963). Whatever the exact cause of the observed specificity of this stain for plasmalemma, the latter authors suggest that staining relates to secretory activity while Roland (1969) somewhat similarly suggests that it relates to the extracellular synthesis of polysaccharide. Neither hypothesis has been proven but a lack of polysaccharide synthesis activity associated with the extrahaustorial membrane might also explain the nongranular nature of this membrane since granules are commonly thought to be involved in the synthesis of at least the fibrillar components of higher-plant cell walls (e.g., Willison and Brown, 1977). Significantly, the extrahaustorial membrane surrounding M-haustoria of *Uromyces phaseoli* var. *vignae* does stain when conventional staining shows the extrahaustorial matrix to contain electron-opaque material (Fig. 54). This matrix material is also strongly PACP positive, and if it represents extracellular polysaccharide synthesis by the host, then this would be consistent with evidence from other types of biotrophic, symbiotic, relationships that the symbiont often becomes surrounded by polysaccharide components of the host wall (Cox and Sanders, 1974; Lalonde and Knowles, 1975; Manocha and Lee, 1972; Manocha and Letourneau, 1978).

PACP staining does not always produce the results described for uredial infections of *Melampsora lini* and pycnial infections of *Uromyces phaseoli* var. *vignae*. Application of this stain to uredial infections of *Uromyces phaseoli* var. *vignae* (M. C. Heath, unpublished) and *Puccinia graminis* f. sp. *tritici* (Harder *et al.*, 1978) has so far resulted in the extrahaustorial membrane showing more consistent, and often greater, staining than the rest of the host plasmalemma (Fig. 62). The specificity of the stain seems to be otherwise normal, at least for *U. phaseoli* var. *vignae*, since the fungal plasmalemma stains strongly whereas other fungal membranes do not. Obviously this variability in response to PACP staining needs further investigation but it may reflect a real change in the nature or synthetic activity of both invaginated and noninvaginated portions of the host plasmalemma with age of infection. Such an explanation is not unlikely since plasmolysis (e.g., Thatcher, 1943) and solute leakage experiments (Elnaghy and Heitefuss, 1976; Hoppe and Heitefuss, 1974) suggest that the general permeability of host cells, and presumably the plasmalemma, increases as infection progresses. There is also some supporting ultrastructural evidence for this since, in advanced infections of *Puccinia sorghi* (Heath, 1976b) or *Uromyces phaseoli* var. *vignae* (uredial or pycnial, M. C. Heath, unpublished) treated with uranyl acetate (which normally is translocated apoplastically and does not cross the plasmalemma), none of the usual deposits occur in the walls of either fungus or host and the host cytoplasm shows unusual features, such as extensive development of rough endoplasmic reticulum and polysomes, which suggest that the salt has entered the cell. Significantly, such indications of a change in host permeability are found in cells apparently lacking haustoria and are *not* found in haustoria-containing cells in regions of younger infections where intercellular mycelium is sparse. This suggests that the change in permeability of the host plasmalemma is not mediated merely by the presence of the haustorium, but is more closely related to the amount, or age, of the intercellular hyphae. Such a hypothesis fits Elnaghy and Heitefuss's (1976) conclusion that the increase in permeability in *U. phaseoli* var. *typica* infections is linked to the gradual increase in the amount of methyl-3,4-dimethoxycinnamate (the urediospore germination self-inhibitor) in infected tissue.

Invasion of the host cell seems initially to cause surprisingly little disorganization of the host protoplast (Coffey, 1975; Coffey *et al.*, 1972b; Heath and Heath, 1971, 1975; Littlefield and Bracker, 1972) and most of the differences in ultrastructure between infected and uninfected cells involve the frequency or distribution of certain components. One of the more striking examples of this phenomenon involves the endoplasmic reticulum (ER) and virtually every study of M- or D-haustoria shows cisternae of ER lying parallel to the extrahaustorial membrane (Fig. 52) (Abu-Zinada *et al.*, 1975; Ehrlich and Ehrlich, 1971a,b; Harder, 1978; Hardwick *et al.*,1971; Heath, 1972; Heath and Heath, 1971, 1975; Kajiwara, 1971; Kohno *et al.*, 1977a; Littlefield and Bracker, 1972; Manocha and Shaw, 1967; Müller *et al.*, 1974a; Oláh *et al.*, 1971; Orcival, 1969; Shaw and Manocha, 1965b; Van Dyke and Hooker, 1969; Zimmer, 1970) and several studies suggest that this association is particularly consistent or obvious around the necks of D-hau-

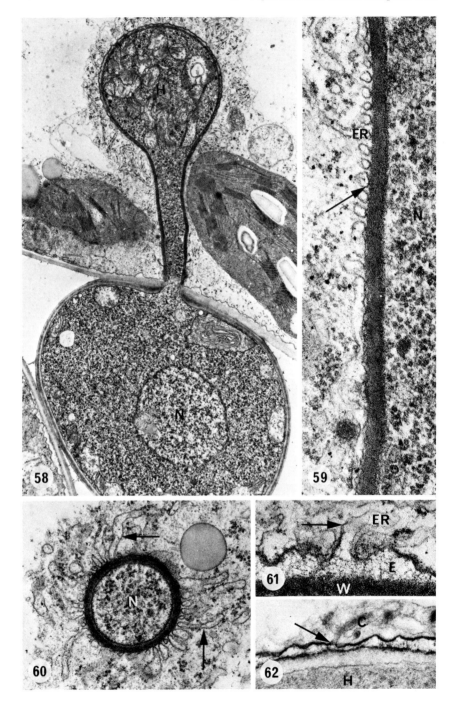

storia (Ehrlich and Ehrlich, 1971a; Hardwick *et al.*, 1971; Heath and Heath, 1971; Littlefield and Bracker, 1970, 1972). When tissue is prepared for electron microscopy by a method that will reveal ribosomes (i.e., not permanganate fixation), the ER around the haustorium usually appears rough (Fig. 52) and the associated ribosomes may show frequent polysome configurations (Hardwick *et al.*, 1971). When cisternae are in close proximity to the extrahaustorial membrane, however, the face closest to this membrane is usually ribosome-free (Fig. 52) (Hardwick *et al.*, 1971; Heath, 1972; Heath and Heath, 1971; Littlefield and Bracker, 1972; Orcival, 1969). Several authors suggest that continuity may exist between this ER and the extrahaustorial membrane (Bossányi and Oláh, 1972; Calonge, 1969; Ehrlich and Ehrlich, 1971b; Kajiwara, 1971; Shaw and Manocha, 1965b) but micrographs used to support this hypothesis commonly lack clear, indisputable, membrane continuity; one of the more convincing examples from *Puccinia graminis* f. sp. *tritici* infections (Harder *et al.*, 1978) is shown in Fig. 61. Bracker and Littlefield (1973) and Heath (1972) point out that such continuity between the plasmalemma and the ER is inherently unlikely in a normal situation because of known structural and chemical differences between these two membranes (Morré, 1975). Nevertheless, the extrahaustorial membrane is not a "normal" plasmalemma (see preceding paragraphs, this section), so this argument may not be valid. Certainly it is important to know whether such connections really exist since conceivably they could be extremely important in facilitating the transfer of host materials to the fungus.

Figs. 58–68. Responses of the host endomembrane system.

Fig. 58. A young D-haustorium (H) of *Puccinia graminis* f. sp. *tritici* [note absence of a neckband (compare with Figs. 22–24) and the presence of a nucleus (N) in the haustorial mother cell]. Profiles of host endoplasmic reticulum can be seen aggregated around the haustorium. (× 12,000) (From Harder *et al.*, 1978. Figures 58–62, reproduced by permission of the National Research Council of Canada from *Can. J. Bot.* **56,** 2955–2966.)

Fig. 59. Detail of the upper part of the neck (N) of the haustorium shown in Fig. 58. Note the regular association of what seems to be endoplasmic reticulum (ER) along the haustorium, and the increase in electron opacity of the ER membrane (arrow) where it touches the extrahaustorial membrane. (× 70,000) (From Harder *et al.*, 1978.)

Fig. 60. Cross section of a haustorial neck (N) similar to that shown in Fig. 58. Note the cisternae of host endoplasmic reticulum (arrows) radiating from the fungus. (× 36,900) (From Harder *et al.*, 1978.)

Fig. 61. A section through the body wall (W) and extrahaustorial matrix (E) of a young D-haustorium of *P. graminis* f. sp. *tritici* showing a possible connection (arrow) between the extrahaustorial membrane and surrounding host endoplasmic reticulum (ER). (× 60,000) (From Harder *et al.*, 1978.)

Fig. 62. A periodic acid–chromic acid–phosphotungstic acid (PACP) treated section of a young D-haustorium (H) of *P. graminis* f. sp. *tritici* showing a possible connection (arrow) between the intensely-stained extrahaustorial membrane and the less intensely-stained membrane complex (C) similar to that shown in Fig. 63. (× 30,000) (From Harder *et al.*, 1978.)

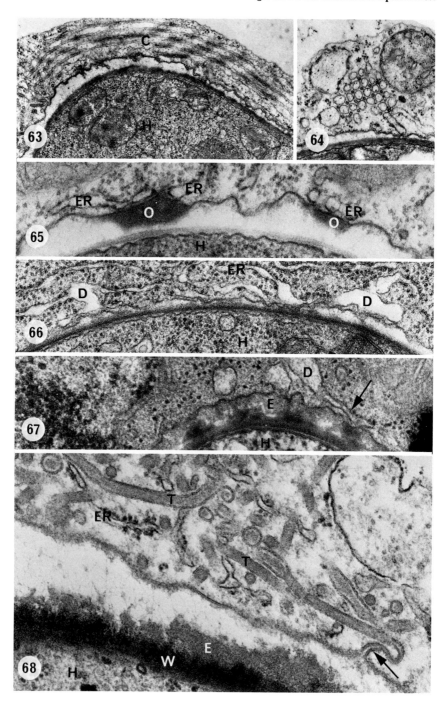

The variability in the arrangement of the ER surrounding the haustorium (Ehrlich and Ehrlich, 1971a; Heath and Heath, 1971) suggests that this association may be in a state of flux. Information from uredial infections of *Puccinia gramminis* f. sp. *tritici* (Harder *et al.*, 1978) and *Uromyces phaseoli* var. *vignae* (Heath, 1972; Heath and Heath, 1971, 1975; M. C. Heath, unpublished) suggests that some of this variability may reflect the different ages of the haustoria examined and that striking changes may occur during the early stages of D-haustorium formation. The association between ER and the D-haustorium seems to develop soon after penetration since cisternae lying parallel to the haustorial neck of *U. phaseoli* var. *vignae* are present before the expansion of the haustorial body (Fig. 32) (Heath and Heath, 1975). In both fungi, the development of the body is accompanied by marked proliferation of ER, particularly around the upper portion of the neck (Figs. 58–60) where some of the cisternae appear to radiate from the haustorium out into the cytoplasm (Harder *et al.*, 1978; M. C. Heath, unpublished). This is most striking in *Puccinia graminis* f. sp. *tritici* (Fig. 60). In Fig. 59 it can be seen that the ER membrane seems to actually contact the extra-

---

Fig. 63. Body (H) of a D-haustorium of *Puccinia graminis* f. sp. *tritici*, older than that shown in Fig. 58. Note the complex of membranes (C) in the surrounding host cytoplasm. (× 6,000) (From Harder *et al.*, 1978. Figures 63 and 64 reproduced by permission of the National Research Council of Canada from *Can. J. Bot.* **56**, 2955–2966.)

Fig. 64. Cross section of a membrane complex similar to that shown in Fig. 63. Note the orderly arrangement of two types of tubules, smaller ones containing electron-opaque material and larger ones with electron-lucent contents. (× 26,000) (From Harder *et al.*, 1978.)

Fig. 65. A young (less than 24 hours old), first-formed D-haustorium (H) of *Uromyces phaseoli* var. *vignae*. Note the electron-opaque material (O) in the extrahaustorial matrix where the host endoplasmic reticulum (ER) comes in close association with the extrahaustorial membrane. (× 63,200) (M. C. Heath, unpublished.)

Fig. 66. Another young, first-formed D-haustorium (H) of *U. phaseoli* var. *vignae*. Note the abundance of rough endoplasmic reticulum (ER) in the host cytoplasm and how some portions (D) adjacent to the extrahaustorial membrane appear dilated and ribosome free. (× 46,900) [From Heath and Heath, 1971. Reproduced with permission from *Physiol. Plant Pathol.* **1**, 277–287. Copyright Academic Press, Inc. (London), Ltd.]

Fig. 67. An M-haustorium (H) of *U. phaseoli* var. *vignae* showing a dilated (D) portion of endoplasmic reticulum (arrow) containing electron-opaque contents somewhat similar to the material within the adjacent extrahaustorial matrix (E). (× 53,500) (M. C. Heath, unpublished.)

Fig. 68. A mature D-haustorium (H) of *Melampsora lini*. Diffuse electron-opaque material (E) can be seen in the extrahaustorial matrix next to the haustorial wall (W), and membrane-bound tubules (T) containing similar material occur in the adjacent host cytoplasm. Note the apparent connection (arrow) between these tubules and the extrahaustorial membrane and the similarity in appearance of the two types of membrane [both of which are distinct from that of the rough endoplasmic reticulum (ER)]. (× 59,000). [Micrograph by M. D. Coffey, from Bracker and Littlefield, 1973. Reproduced with permission from "Fungal Pathogenicity and the Plant's Response" (R. J. W. Byrde and C. V. Cutting, eds.). Copyright Academic Press, Inc. (London), Ltd.]

haustorial membrane and the former appears to stain more strongly where the two membranes touch; however, continuity between the two membranes has never been observed at this stage, although profiles suggesting such continuity are often seen around the haustorium body (Fig. 61) (Harder *et al.*, 1978). Around this latter structure, complexes of large and small, seemingly interconnected, ribosome-free tubules develop which are possibly connected to, and derived from, the surrounding rough ER. While the proliferation of ER around the haustorial neck becomes less marked as the haustorium matures, these complexes become more extensive (Figs. 63 and 64). Both large and small tubules often seem connected to the extrahaustorial membrane although only the latter tubules have a limiting membrane which resembles the extrahaustorial membrane in thickness and staining (the larger tubules have membranes resembling those of the ER); PACP treatment stains only the extrahaustorial membrane (Fig. 62). Somewhat similar complexes have been reported around haustoria of the downy mildew *Peronospora pisi* (Hickey and Coffey, 1977) and the smaller tubules resemble the tentaclelike invaginations of the extrahaustorial membrane reported previously around young haustoria of *P. graminis* f. sp. *tritici* (Ehrlich and Ehrlich, 1971a) and older haustoria of *Melampsora lini* and *P. helianthi* (Fig. 68) (Coffey *et al.*, 1972a). In the latter situation their appearance coincides with the development of what was later (Coffey, 1976) considered to be electron-opaque components of the extrahaustorial matrix. Whether these "tubules" have a role in the synthesis of matrix material is a matter of speculation, but both matrix and tubule contents have similar high electron-opacities.

While membrane complexes and tentaclelike extensions of the extrahaustorial membrane have not been observed in uredial *Uromyces phaseoli* var. *vignae* infections, around young (less than 12 hours old) haustoria, portions of parallel-oriented ER close to the extrahaustorial membrane become associated with regions of strong electron-opacity in the extrahaustorial matrix (Fig. 65). This association has been seen in two susceptible, and one resistant, cultivars (Heath, 1972; M. C. Heath, unpublished) and therefore seems to be a general phenomenon. However, since it has been seen only around the first-formed haustorium at a given infection site, it is as yet unclear whether it occurs in haustoria of a similar age in more established infections. Young haustorial bodies of this rust are sometimes also surrounded by dilated, ribosome-free, regions of ER cisternae containing material similar in electron-lucency to the adjacent extrahaustorial matrix (Fig. 66) (Heath and Heath, 1971). A similar phenomenon has also been seen around M-haustoria of this species where both extrahaustorial matrix and lumen of the dilated portion of the ER contain similar electron-opaque material (Fig. 67). These observations could be interpreted as indicating that ER-derived products are released into the matrix (Heath and Heath, 1971) but the validity of this interpretation remains to be established.

The picture that emerges from these studies is that a marked effect on the host ER system seems to be a characteristic response of susceptible plants to M- and D-haustorium formation although details of the response may depend on the host species. For D-haustoria, there is evidence that the ER may play different roles at different stages of invasion. Presumably this role is the synthesis of, and/or provision of the fungus with whatever components the haustorium needs from the host protoplast. However, Mendgen (1975) has shown peroxidase activity associated with the ER, Golgi bodies, and the nuclear envelope in the vicinity of the D-haustoria of *Uromyces phaseoli* var. *typica*, a phenomenon difficult to interpret in terms of fungal nutrition. A point of interest is that the Golgi apparatus does not seem to be involved in any of the described associations of the ER and the extrahaustorial membrane, in spite of its normal intermediary role in the transformation of ER-type to plasmalemma-type membranes in normal cells (Morré, 1975). Indeed, if Harder's interpretation is correct (see above this section), such membrane transformation occurs directly in the tubular complexes seen around the haustorium body of *Puccinia graminis* f. sp. *tritici*, and a similar phenomenon is suggested to occur during rapid encasement formation in *Uromyces phaseoli* var. *vignae* infections (see Chapter 6, Section I,B,1 and also Section II,C,2, this chapter). In terms of present concepts, therefore, the effect of invasion of the host cell on the endomembrane system seems pronounced and results in situations not normally found in uninvaded host cells. Obviously this is a fruitful area for further study.

Not only does the Golgi apparatus seem not to be involved in the association of the ER with the haustorium, but an increase in the number of Golgi bodies in response to infection has been reported in relatively few rust infections (Ehrlich and Ehrlich, 1971b; Manocha and Shaw, 1967; Orcival, 1969; Pinon *et al.*, 1972; Shaw and Manocha, 1965b; Van Dyke and Hooker, 1969). Cytoplasmic vesicles in the vicinity of the haustorium have been occasionally reported (Abu-Zinada *et al.*, 1975; Ehrlich and Ehrlich, 1963; Manocha and Shaw, 1967; Orcival, 1969; Pinon *et al.*, 1972; Shaw and Manocha, 1965b), but as Ehrlich and Ehrlich (1971b) point out there is little evidence to support the common suggestion that these vesicles are either Golgi derived or that they represent stages in transport of materials to or from the haustorium.

Micrographs of both M- and D-haustoria often show the close proximity of host mitochondria, chloroplasts, and other organelles (Fig. 22). In the absence of any quantitative data, the significance of such observations is difficult to judge, and even if these associations are real, they may not indicate any functional interrelationship (Bracker and Littlefield, 1973). The suggestion that there is direct continuity between the membranes of the chloroplast and the extrahaustorial membrane (Ehrlich and Ehrlich, 1971b) is difficult to believe from micrographs where many organelle membranes appear disorganized and ruptured (Bracker and Littlefield, 1973) but as the latter authors point out, such claims deserve further investigation as to their validity.

In contrast to the uncertainty as to whether the above-mentioned organelles preferentially associate with the haustorium, the close association of the haustorium and the host nucleus has been reported enough times from both electron microscopic (Coffey, 1975; Coffey *et al.*, 1972a; Heath and Heath, 1971; Manocha and Shaw, 1966; Oláh *et al.*, 1971; Van Dyke and Hooker, 1969) and light microscopic (R. F. Allen, 1928, 1932b; Colley, 1918; Rice, 1927) studies to suggest this may indeed be a characteristic response to invasion by both M- and D-haustoria. The indentation of the nucleus by a haustorial lobe seems particularly common in cells invaded by M-haustoria (Figs. 52 and 80) (*Cronartium ribicola*, Colley, 1918; *Puccinia recondita*, Gold *et al.*, 1979; *Melampsora pinitorqua*, M. D. Coffey, personal communication; *Uromyces phaseoli* var. *vignae*, M. C. Heath, unpublished). The apparent lack of any association with host nuclei in other ultrastructural studies may be a chance result of unfavorable planes of section (Bracker and Littlefield, 1973); another possibility, however, is that the association may only exist during the early stages of haustorium formation as suggested for D-haustoria of *U. phaseoli* var. *vignae* (Heath and Heath, 1971).

Only a few studies, primarily of D-haustoria, have examined other aspects of the host response in detail and these suggest that, apart from the changes in frequency and distribution of organelles described above and an increase in starch grains (Coffey *et al.*, 1972b), few major ultrastructural modifications occur until late in the infection process (Figs. 22, 69, and 72), often only after the onset of sporulation (Abu-Zinada *et al.*, 1975; Coffey *et al.*, 1972b; Heath, 1974b; Manocha and Shaw, 1966). Most frequently, the observed response of the host tissue at this stage of infection is a general disorganization of the cytoplasm and a disruption of cell membranes (Coffey *et al.*, 1972b; Ehrlich and Ehrlich, 1963; Manocha and Shaw, 1966; Moore and McAlear, 1961; Shaw and Manocha, 1965b; Van Dyke and Hooker, 1969; Zimmer, 1970), particularly in the center of the infected area (Coffey *et al.*, 1972b; Shaw and Manocha, 1965b). This disorganization is not necessarily accompanied by a similar disorganization of the haustorium (Ehrlich and Ehrlich, 1971a) and in this respect it resembles the cell necrosis observed in some intermediate types of cultivar resistance (see Chapter 6, Section I,B,1).

Detailed observations of the changes which occur in the host areas where complete disruption does not take place are rare, but an increase in the density of the interchromatin regions of the host nucleus has been reported for uredial infections of *Puccinia graminis* f. sp. *tritici* (Manocha and Shaw, 1966) and *Uromyces fabae* (Abu-Zinada *et al.*, 1975). In the former species, this seems to correlate with the increase in RNA known to occur at this stage of infection, and is followed by the eventual disappearance of the chromatin (Manocha and Shaw, 1966). Somewhat similar changes in the nuclei also occur in tissue cultures of *Pinus monticola* in the vicinity of thalli of *Cronartium ribicola* (Robb *et al.*, 1975a). The most detailed and comprehensive study of the effect of infection on mitochondria and microbodies is that of Coffey *et al.* (1972b). In peripheral regions of uredial infections of *Puccinia helianthi*, but not *Melampsora lini*, atypical platelike cristae and occasional crystals develop in the mitochondria (Fig. 70) while, in both types of infection, the microbodies more frequently contain crystalline cores (Fig. 71). The significance of these observations, however, is unknown.

In comparison to nuclei, mitochondria, and microbodies, chloroplast structure in rust-infected tissues has been studied more often, but again primarily in uredial and telial infections (*Uromyces fabae*, Abu-Zinada *et al.*, 1975; *Melampsora lini* and *Puccinia helianthi*, Coffey *et al.*, 1972b; *P. graminis* f. sp. *tritici*, Thomas and Isaac, 1967; *Uromyces phaseoli* var. *vignae*, Heath, 1974b; *M. pinitorqua*, Mlodzianowski and Siwecki, 1975). In all these examples, the chloroplasts eventually show disorganization of the granal structure and/or a reduction in the number of photosynthetic lamellae (Fig. 73). An increase in the development of a peripheral reticulum may also occur (Coffey *et al.*, 1972b; Abu-Zinada *et al.*, 1975; Heath, 1974b) as well as a marked increase in the number and/or size of the plastoglobuli (Abu-Zinada *et al.*, 1975; Coffey *et al.*, 1972b; Heath, 1974b; Mlodizanowski and Siwecki, 1975; Thomas and Isaac, 1967). Prolamellar bodies and electron-opaque structures resembling carotenoid crystalloids occur in the chloroplasts near the sporulating regions of *Uromyces phaseoli* var. *vignae* infections (Fig. 73) (Heath, 1974b) while phytoferritinlike granules often occur in similarly situated chloroplasts in *M. lini* infections (Coffey *et al.*, 1972b). Apparent degradation of starch to electron-opaque granules resembling glycogen (see Chapter 6, Fig. 6) occurs in both *M. lini* (Coffey *et al.*, 1972b) and *M. pinitorqua* (Mlodzianowski and Siwecki, 1975) infections but starch grains appear numerous and undegraded in infections of *U. phaseoli* var. *vignae* of equivalent age (Heath, 1974b).

Interestingly, none of these changes in chloroplast ultrastructure have been mentioned in published studies of basidiospore-derived infections (except an increase in the number of plastoglobuli in tissue culture cells adjacent to the thalli of *Cronartium ribicola*, Robb *et al.*, 1975a). Such changes have been looked for particularly in pycnial infections of *Uromyces phaseoli* var. *vignae* and have not been found even when the pycnia are several weeks old (M. C. Heath, unpublished). Similarly, none of these changes seem to occur in the basidiospore-derived, systemic pycnial infections of *Endophyllum euphorbiae-sylvaticae* (Orcival, 1972) but in this case, the

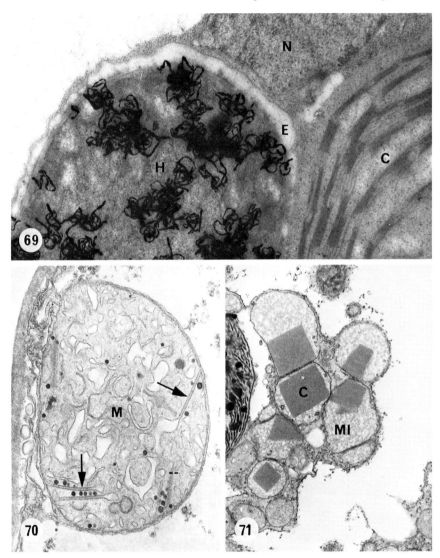

Figs. 69–75.   Responses of host organelles.

Fig. 69.   Autoradiograph of the first-formed D-haustorium (H) of *Uromyces phaseoli* var. *typica* derived from tritium labeled urediospores. Note the close association of the host nucleus (N) and chloroplast (C) with the haustorium. At this early stage of infection, no silver grains can be found in the extrahaustorial matrix (E) or the host cytoplasm, in spite of the heavy labeling of the haustorium. (× 16,000) (From Mendgen and Heitefuss, 1975.)

Fig. 70.   Host mitochondrion (M) in a sporulating uredial infection of *Puccinia helianthi*. Note the atypical, platelike cristae (arrows) and small electron-opaque granules. (× 28,000) (From Coffey *et al.*, 1972a. Figures 70 and 71 reproduced by permission of the National Research Council of Canada from *Can. J. Bot.* **50**, 231–240.)

presence of the fungus during leaf development seems to delay the maturation of the proplastids, and the mature chloroplasts, when formed, show precocious dilation of the thylakoids and disruption of the plastid envelope.

Information concerning ultrastructural changes in components other than the major host organelles is lacking in many rust infections although in host cells adjacent to *Endophyllum euphorbiae-sylvaticae* (Orcival, 1972), electron-opaque globules develop near the cell wall, and on various membranous components of the cell, in a similar manner to that observed in the intermediate type of cultivar resistance to *Uromyces phaseoli* var. *vignae* (see Chapter 6, Section I,B,1, and Fig. 14). Lipid bodies increase in the cytoplasm of tissue culture cells of *Pinus monticola* infected with *Cronartium ribicola* and tiny electron-opaque deposits develop uniformly throughout the cytoplasm in cells 1–2 mm from the fungal thallus (Robb *et al.*, 1975a). These deposits resemble the "pepper" of unknown origin which many electron microscopists occasionally observe in their micrographs. However, in *P. monticola,* they are suggested to represent regions of tannin deposition.

Several authors have considered these infection-induced changes in ultrastructure of various organelles to be a sign of premature senescence of the host cell (Coffey *et al.*, 1972b; Orcival, 1972; Shaw and Manocha, 1965b) and, indeed, in *Puccinia graminis* f. sp. *tritici* infections, these changes resemble those of detached, senescing, uninfected host leaves (Shaw and Manocha, 1965a, 1965b). However, the changes in chloroplast ultrastructure in the center of *Uromyces phaseoli* var. *vignae* infections do *not* resemble those seen in uninfected leaves allowed to senesce on the intact plant (compare Figs. 73 and 74) but changes similar to those seen in infected tissue can be induced in uninfected leaves by ethylene treatment (Fig. 75) (Heath, 1974b). In fact, many of the alterations in chloroplast structure reported for rust infections resemble ethylene-induced chromoplast development in ripening fruit (Coffey *et al.*, 1972b; Heath, 1974b). Since susceptible leaves infected with *U. phaseoli* var. *vignae* release more ethylene than noninfected or senescent leaves, it seems likely that the increased concentration of this gas in infected areas is primarily responsible for many of the chromoplastlike changes which take place in the chloroplast. Whether these changes resemble those of senescing tissue probably depends on the relative degrees of ethylene production during the two processes (Heath, 1974b).

Why chromoplastlike chloroplasts are not found in pycnial infections, particularly those of *U. phaseoli* var. *vignae* where they occur in the same host in uredial and telial stages, is an interesting question. Perhaps the answer lies in

---

Fig. 71. Host microbodies (MI) in a sporulating uredial infection of *Melampsora lini.* Note the unusually high frequency of crystalline inclusions (C) in these microbodies. (× 15,000) (From Coffey *et al.*, 1972a.)

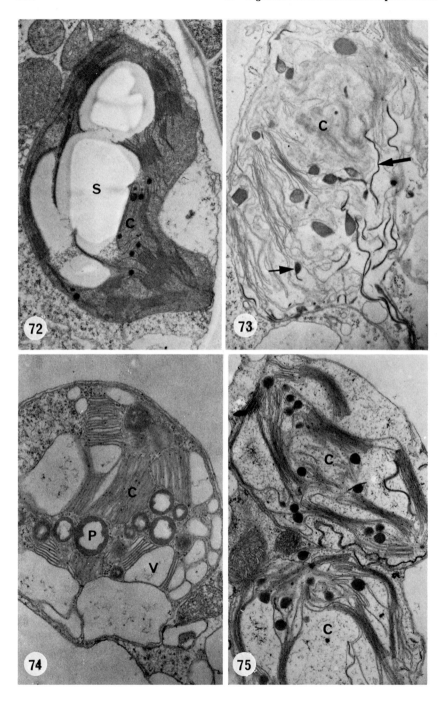

the fact that sporulation in the last two stages represents the end of both vegetative growth and the maintenance of compatibility with the host, whereas the "health" of the host must still be maintained after pycnium formation. More ultrastructural work is needed, however, to determine whether this difference between pycnial and uredial infections is a general phenomenon.

One final point is that it is generally unknown, or unreported, whether the changes in organelle structure which occur after sporulation take place in noninvaded as well as invaded host cells. The lack of a haustorium in a particular cell is extremely difficult to prove without large numbers of serial sections but the general impression from published and unpublished work is that both noninvaded, as well as invaded, cells are involved. However, the validity of this conclusion needs to be substantiated.

2. Collars and Encasements

A common response of the host cell to the formation of both M- and D-haustoria is the deposition of material against the host cell wall. Although such wall appositions have been reported in regions of invaded cells away from the penetration site of the haustorium (Ehrlich et al., 1968b), such situations seem relatively uncommon and more frequently this type of host response results in the haustorium being surrounded by a "collar" of material at the point where it enters the host cell (Abu-Zinada et al., 1975; Calonge, 1969; Coffey, 1975; Coffey et al., 1972a; Ehrlich et al., 1968b; Harder, 1978; Hardwick et al., 1971; Heath, 1972; Heath and Heath, 1971; Kajiwara, 1971; Kohno et al., 1977a; Littlefield and Bracker, 1972; Müller et al., 1974a; Rijkenberg and Truter, 1973a; Rijo and Sargent, 1974; Robb et al., 1975b; Shaw and Manocha, 1965b; Van Dyke and Hooker, 1969; Walles,

---

Fig. 72. Typical starch-containing (S) chloroplast (C) of *Vigna sinensis* in a nonsporulating uredial infection of *Uromyces phaseoli* var. *vignae.* (× 18,400) (From Heath, 1974b. Figures 72–75, reproduced by permission of the National Research Council of Canada from *Can. J. Bot.* 52, 2591–2597.)

Fig. 73. Typical chloroplast (C) from *Vigna sinensis* cells closest to the zone of developing urediospores in a *U. phaseoli* var. *vignae* sporulating pustule. The normal thylakoid arrangement is disrupted (compare with Fig. 72) and plastoglobuli are often seen associated with membrane fragments (small arrow). Note the electron-opaque sheets (large arrow), resembling the carotenoid crystalloids of certain chromoplasts, which (in other sections) seem to be derived by deposition of material between adjacent thylakoid pairs. (× 18,400) (From Heath, 1974b.)

Fig. 74. A chloroplast (C) from an uninfected, naturally senescing, yellowing *Vigna sinensis* leaf. The thylakoids have dilated to form vesicles (V) and the plastoglobuli are unusually large and have electron-lucent centers (P). Note the differences between this chloroplast and that in sporulating, infected, tissue (Fig. 73). (× 30,400) (From Heath, 1974b.)

Fig. 75. Two chloroplasts (C) in an uninfected, ethylene treated, yellowing *Vigna sinensis* leaf. Note the similarity in appearance between these and the chloroplasts found in sporulating *U. phaseoli* var. *vignae* infections (Fig. 73). (× 16,200) (From Heath, 1974b.)

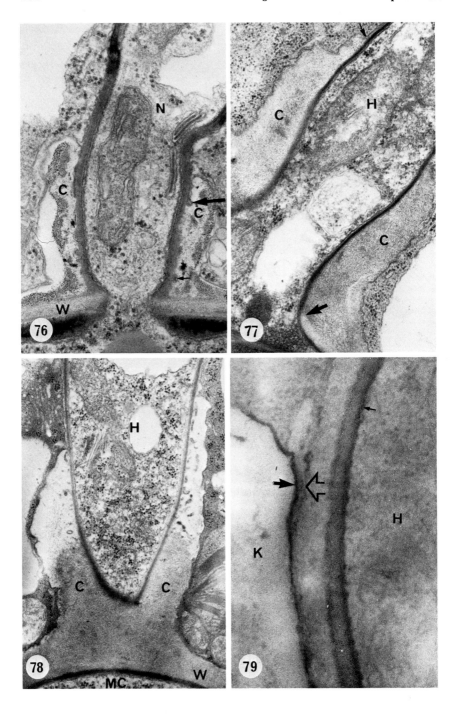

1974). These collars are always in contact with the host cell wall but may, or may not, resemble it in appearance. In cases where the collar is primarily electron-lucent, the boundary with the wall is easily discerned (Fig. 76) (e.g., Coffey *et al.*, 1972a; Ehrlich *et al.*, 1968b; Littlefield and Bracker, 1972) suggesting the collar is not merely wall material displaced inward during haustorium penetration. These translucent collars usually contain membranous vesicles (such as shown in Chapter 6, Fig. 23) (e.g., Coffey *et al.*, 1972a; Ehrlich *et al.*, 1968b; Littlefield and Bracker, 1972) or more extensive electron-opaque patches (Ehrlich *et al.*, 1968b; Rijkenberg and Truter, 1973a) but such inclusions are usually lacking from collars with a more fibrillar appearance (Fig. 77). The boundary between these fibrillar collars and the cell wall may be less distinct and sometimes indiscernible (Fig. 78) (Abu-Zinada *et al.*, 1975; Harder, 1978; Hardwick *et al.*, 1971; Orcival, 1969; Plotnikova *et al.*, 1977; Walles, 1974). Thus, in these cases it cannot be ruled out that the collars contain original wall material pushed inward dur-

---

Figs. 76–81. Collars and encasements.

Fig. 76. A neck (N) of a D-haustorium of *Puccinia sorghi*. Collar (C) material is present around that portion of the neck closest to the site of penetration. The small electron-opaque granules in the collar seem to be an artifact caused by the prefixation treatment of this material with uranyl acetate, but they clearly differentiate the collar from the host cell wall (W). Note the membrane fragments (small arrow) within the more translucent portions of the collar and the way in which the host plasmalemma is folded into the collar for a short distance (large arrow). This situation may represent the development of a Type I collar from a Type II (see Section II,C,2, this chapter) but it is possible that the folded plasmalemma may have originally extended farther into the collar and that subsequent membrane fusion and disintegration have taken place. (For a typical Type I collar see Chapter 6, Fig. 23.) (× 45,600)(M. C. Heath, unpublished.)

Fig. 77. A fibrillar Type II collar (C) (for diagram see Fig. 20, this chapter) around an M-haustorium (H) of *Uromyces phaseoli* var. *vignae.* The termination of the host wall at the site of penetration can be clearly seen (large arrow) and the morphological distinction between wall and collar suggests that they are not identical in composition. Note that the invaginated host plasmalemma does not line the inside of the collar but continues directly along the haustorium after the collar terminates (small arrow). (× 28,800) (M. C. Heath, unpublished.)

Fig. 78. An oblique section through the penetration site of an M-haustorium of *U. phaseoli* var. *vignae;* the connection between the haustorium (H) and the mother cell (MC) is in an adjacent section. Note the fibrillar nature of the lower part of the Type II collar (C) and the primarily electron-lucent appearance of the upper portion. In this example, there is little clear distinction between the collar and the host cell wall (W). (× 24,200) (M. C. Heath, unpublished.)

Fig. 79. A periodic acid–chromic acid–phosphotungstic acid (PACP) treated section of an M-haustorium (H) of *U. phaseoli* var. *vignae* partially encased in callose-like electron-lucent material (K) such as shown in Figs. 80 and 81. The fungal plasmalemma (small solid arrow) and the host plasmalemma covering the encasement (large solid arrow) stain intensely while the extrahaustorial membrane (open arrow) does not. Note that the inside of the encasement, like a Type I collar, is lined by the host plasmalemma (large solid arrow) and that the fungus is therefore separated from the encasement by two portions of this host membrane (the encasement membrane and the extrahaustorial membrane). (× 74,300) (M. C. Heath, unpublished.)

ing haustorium penetration; however, the volume of the collar is commonly too large to be completely accounted for in this way.

Collars containing both fibrillar and electron-lucent components have been observed (Fig. 78) (Coffey *et al.*, 1972a; Hardwick *et al.*, 1971; Kajiwara, 1971; Littlefield and Bracker, 1972) and primarily translucent, or primarily fibrillar, collars may occur in single rust infections (Coffey *et al.*, 1972a; Hardwick *et al.*, 1971). Thus it seems that not all collars are of the same chemical nature, nor is each one necessarily homogeneous. In susceptible plants infected with *Uromyces phaseoli* var. *typica* (Hardwick *et al.*, 1971), and in resistant host cultivars infected with *U. phaseoli* var. *vignae* (Heath, 1971), light microscopy has shown some collars to fluoresce under ultraviolet light following aniline blue treatment. This suggests that collars can contain callose and, since this polysaccharide is typically electron-lucent (e.g., Heslop-Harrison, 1966), it seems likely that the more electron-lucent collars have this carbohydrate as a major component (although there is some evidence from resistant interactions that other polysaccharides may be present, see Chapter 6, Section I,B,1). Heath and Heath (1971) suggest that the membranous material often found in these collars represents excess membrane trapped when vesicles containing collar material liberate their contents by fusion with the host plasmalemma. The origin of these vesicles is unknown but Littlefield and Bracker (1972) suggest that they are derived from the Golgi apparatus. However their micrographs show both Golgi vesicles and dilated endoplasmic reticulum (ER) in the vicinity of a developing collar, and dilated ER has been implicated to have a role in the development of the more extensive, but otherwise similar, collars and encasements found in certain resistant interactions (Heath and Heath, 1971; see Chapter 6, Section I,B,1). The absence of trapped membranes in the primarily fibrillar collars suggests that the fibrillar components may not be deposited in the same manner as the electron-lucent materials. Whether these fibrillar collars contain materials similar to the host cell wall, as their appearance suggests, is unknown.

Whatever the source, or nature, of the collar components, there appear to be two types of collars found in rust infections. The more electron-lucent collars commonly appear to "grow up" from the host wall during (Littlefield and Bracker, 1972), or possibly after, fungal penetration in such a manner that they are surrounded inside and out by the host plasmalemma (Fig. 20a,b) (Coffey *et al.*, 1972a; Harder, 1978; Littlefield and Bracker, 1972; Kohno *et al.*, 1977a). Thus the collar is separated from the fungus by both this membrane and the extrahaustorial membrane (Type I collar, Fig. 20a, also see similar collar in Chapter 6, Fig. 23). In contrast, a survey of published micrographs shows that the more fibrillar collars are often *not* separated from the haustorium by any discernible membranes (Type II collar, Fig. 20c). Harder (1978) and Walles (1974) explain this observation on the hypothesis that such collars are formed as the fungus grows through material deposited either on the cell wall before penetration, or directly over the surface of the young haustorium as it enters the host cell (Fig. 20d). In support of this hypothesis, small (and therefore possibly young) haustoria of *Melampsora pinitorqua* have been seen completely encased in fibrillar, collarlike material deposited next to the haustorial wall (M. D. Coffey, personal communication). [However other examples of similarly encased young haustoria (Hardwick *et al.*, 1971) where serial sections were not examined must be viewed with some skepticism since similar-appearing micrographs can result from oblique sectioning through collars of more developed haustoria.]

The available ultrastructural evidence suggests that well developed fibrillar, Type II collars are frequently found around M-haustoria (Fig. 77) (*Cronartium ribicola*, Robb *et al.*, 1975b; *Melampsora pinitorqua*, M. D. Coffey, personal communication; *Peridermium pini*, Walles, 1974; *Puccinia coronata* f. sp. *avenae*, Harder, 1978; *P. sorghi*, Rijkenberg and Truter, 1973a; *Uromyces phaseoli* var. *vignae*, M. C. Heath, unpublished). Electron-lucent collars of some D-haustoria have also been suggested to develop in a similar manner to Type II collars (Kajiwara, 1971; Rijo and Sargent, 1974) and the latter authors suggest that the occasional outward displacement of the host cell wall at the site of penetration may be the result of the force applied by the fungus to break through the papilla of collar material. However neither study provides any evidence that the double membrane which typically lines Type I collars is absent and the (admittedly few) studies of D-haustorium formation have revealed no signs of extensive papillae formation prior to host wall penetration. More convincing evidence of Type II collars around D-haustoria are rare (possibly *Melampsora lini* in Fig. 4 of Coffey *et al.*, 1972a; *Uromyces phaseoli* var. *typica*, Hardwick *et al.*, 1971) although it is possible that what has been interpreted as extrahaustorial matrix in this region (e.g., Fig. 24, see The Haustorial Neck, Section II,A,1,c, this chapter) may, in fact, be more equivalent in mode of origin to a rudimentary collar of this type. In contrast, electron-lucent, Type I collars have been observed around both M- (Kohno *et al.*, 1977a) and D- (Littlefield and Bracker, 1972) haustoria in susceptible hosts but they are not universally present and, in some cases, seem to occur more frequently in older infections (Coffey *et al.*, 1972a; Heath and Heath, 1971). Considering their different modes of origin, it is theoretically possible that a single haustorium may have both a Type I and a Type II collar surrounding it, but clear examples of this have not been reported. However, a Type I collar may seemingly develop by localized extensions of a Type II collar since cases can be found where the folded plasmalemma extends for only a limited distance inside the distal portion of the collar (Fig. 76) (Hardwick et al., 1971).

Normally neither Type I nor Type II collars cover more than a small portion of the haustorium, but more extensive collars have been observed around M- and D-haustoria in susceptible infections and in some instances the haustorium becomes completely encased in collar material. Most frequently, these encasements are primarily electron-lucent, contain membranous profiles, and seem to be extensions of the Type I collar (Ehrlich and Ehrlich, 1971b; Ehrlich et al., 1968b; Heath and Heath, 1971; Kajiwara, 1971; Kohno et al., 1977a; Littlefield and Bracker, 1972; Mims and Glidewell, 1978; Orcival, 1969) since, where this feature has been looked for, they are separated from the haustorium wall by two portions of the host plasmalemma (Heath and Heath, 1971; Littlefield and Bracker, 1972; Kohno et al., 1977a); these membranes become isolated from the host cell once the encasement is complete. Ultrastructural features of encasement formation have not been described in compatible interactions but there is no reason to suppose the process to be different from that described for similar, callose-containing encasements in resistant plants (see Chapter 6, Section I,B,1), although Heath and Heath (1971) suggest that encasement may occur more slowly in plants susceptible to the uredial stage of Uromyces phaseoli var. vignae since fewer membranes seem to be trapped in the encasement material. Periodic acid–chromic acid–phosphotungstic acid (PACP) staining of partially encased M-haustoria of this rust (in susceptible plants) shows that, in contrast to the extrahaustorial membrane, the encasement membrane stains strongly even in the region where one would expect new encasement material to be deposited (Fig. 79) (M. C. Heath, unpublished). This observation supports the conclusion from the study of resistant plants that transformation of the membrane occurs before the vesicles containing encasement precursors fuse with the encasement membrane.

In addition to encasements obviously derived from the extension of Type I collars, M-haustoria of both Kuehneola japonica (Kohno et al., 1977a) and Uromyces phaseoli var. vignae (M. C. Heath, unpublished) can also be encased in similar material but without the double membrane between the encasement and the haustorium wall (Fig. 80). Kohno et al. (1977a) suggest that these encasements form by the deposition of encasement material directly into the extrahaustorial matrix. However, close examination of such encasements around U. phaseoli var. vignae haustoria often reveals fragments of double membranes, or rows of small vesicles, where the double membrane is expected to be (Fig. 81). Thus, it seems possible that the encasements develop normally (i.e., as extensive Type I collars) but the trapped portions of the host plasmalemma and extrahaustorial membrane either disintegrate or fuse in places to form series of vesicular profiles.

The formation of partial or complete, electron-lucent encasements seems to develop frequently in response to the initial invasion of epidermal cells after inoculation with basidiospores of *Kuehneola japonica* (Kohno *et al.*, 1977a) and is also common in epidermal cells invaded at a later stage of infection by M-haustoria of *Uromyces phaseoli* var. *vignae* (M. C. Heath, unpublished). However, M- and D-haustoria in susceptible mesophyll cells are encased relatively infrequently, and usually these encased haustoria are necrotic (Ehrlich and Ehrlich, 1971b; Ehrlich *et al.*, 1968b; Littlefield and Bracker, 1972; Mims and Glidewell, 1978; Orcival, 1969) although this is not always the case (Heath and Heath, 1971; Kohno *et al.*, 1977a). Heath and Heath (1971) have commented that since callose deposition is a frequent, apparently nonspecific response of higher plants to injury, the formation of callose-containing collars and encasements may similarly be a nonspecific response to invasion of the host cell. The fact that they do not normally develop in direct contact with the haustorial surface, but are separated by two layers of the host plasmalemma, suggests that they are not a direct response to the haustorium itself but are induced by some other feature such as the hole in the host cell wall. Indeed, perhaps the most surprising feature of susceptible interactions is that callose synthesis is not induced in every invaded cell, and this may indicate a suppression of this response in compatible associations as long as the haustorium remains alive and unimpaired (Heath and Heath, 1971; Heath, 1974a). Such a hypothesis may explain why collars and encasements are often more frequent in older infections (Coffey *et al.*, 1972a; Heath and Heath, 1971) since the degree of compatibility between host and fungus, and the health of the latter, may decrease with time, particularly after the onset of sporulation.

In contrast to the widespread reports of electron-lucent encasements in susceptible rust infections, fibrillar encasements have so far been illustrated only for M-haustoria of *Peridermium pini* (Walles, 1974) and, possibly (see Section II,B, this chapter) *Puccinia sorghi* (Rijkenberg and Truter, 1973a). In both studies, no membranes exist between these encasements and the haustorial wall, and Walles (1974) suggests that they arise by the synthesis of the collar material (Type II) "keeping pace" with haustorium development.

D. Concluding Remarks

Since the ultrastructural studies described in this section show clearly that M- and D-haustoria differ in a number of ways, the questions which spring to mind are (1) do these structures differ in function, and (2) if they do not, why do such structural differences exist? Unfortunately, although evidence is available for the powdery mildews that the haustorium is an organ of nutrition (Bushnell, 1972; Gay and Manners, 1977; Manners and Gay, 1978), no information exists for either M- or D-haustoria of the rusts concerning the

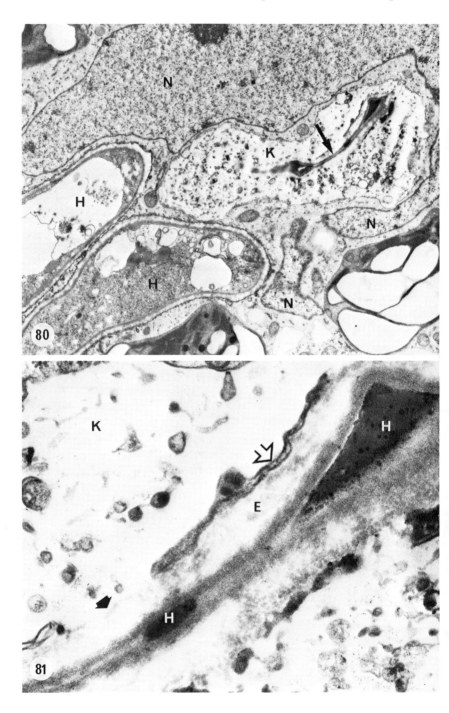

nature of the interaction between the haustorium and the host cell. Autoradiography at the electron microscope level has shown that substances can pass from *Puccinia graminis* f. sp. *tritici* to the host in established infections (Ehrlich and Ehrlich, 1970). In contrast no such transfer could be detected during early stages of *Uromyces phaseoli* var. *typica* infection (Fig. 69) (Mendgen and Heitefuss, 1975) but it is possible that this may have been due to the transferred materials being extracted during specimen preparation since they may not have had time to be incorporated into the nonextractable components of the host cells. Transfer of materials from the host to the fungus has also been demonstrated for these same two rust species (see Chapter 6, Fig. 9) (Favali and Marte, 1973; Manocha, 1975; Mendgen, 1977) but whether either type of transport occurs through the haustorium is unknown. Indeed, in uredial infections, the uptake of labeled glucose by the fungus before haustorium formation (Onoe *et al.*, 1973), the reported growth of intercellular mycelium for 5 days in the absence of haustoria (Pady, 1935a), the change in wall structure as the fungus enters the leaf (see Chapter 3, Section III,A), the general increase in permeability of host cells during fungal growth (see Section II,C,1, this chapter), all imply that the intercellular mycelium obtains at least some nutrients directly from the host cells without passage through the haustorium. If this is so, the D-haustorium may have a more specific role than that of extracting basic nutrients from the invaded cell. There is some evidence that D-haustoria can take up more complicated substances than the infection hypha (Onoe *et al.*, 1973), and it is conceivable that the haustorium mediates the interaction with the host (perhaps of a type suggested by Chakravorty and Shaw, 1977) which removes the metabolic block restricting fungal growth in the absence of a living plant.

The structural differences between D- and M-haustoria, and the closer resemblance of the latter to intercellular hyphae, suggest that their role in the infection process may not be the same. It is also possible that the more determinate types of M-haustoria have a different function from the primarily intracellular mycelium of other species, although Fullerton (1970)

Fig. 80. Profiles of two healthy (H) and one necrotic (arrow) M-haustoria of *Uromyces phaseoli* var. *vignae*. The necrotic haustorium has collapsed and is encased in electron-lucent material (K) containing membranous material. Note the association of these haustoria with the deeply indented host nucleus (N). ($\times$ 8,600) (M. C. Heath, unpublished.)

Fig. 81. A detail of the necrotic haustorium (H) shown in Fig. 80. In some regions (closed arrow), the calloselike encasement material seems to be in direct contact with the extrahaustorial matrix (E). However, in other areas the encasement is separated from the haustorium by lengths of double membranes (open arrow) suggesting that such separation may have originally existed over the whole haustorium (as in Fig. 79) but that the separating membranes subsequently fused and/or disintegrated. Note the membrane bound inclusions in the encasement (k) material, which seem to be characteristic of callose-containing deposits (see also Chapter 6, Figs. 26, 27, 29). ($\times$ 53,400) (M.C. Heath, unpublished.)

suggests for the similar situation which exists in the smuts, that such differences in morphology reflect differences in degree of development rather than differences in kind. Obviously, the roles of all these types of intracellular structures need urgent investigation.

Why such differences in morphology, and possibly function, should exist between M- and D-haustoria of the same rust is an interesting question, particularly in rusts where both structures occur in the same host plant. In fact, the available ultrastructural evidence suggests there to be more similarity between D-haustoria of different rust species than between M- and D-haustoria of the same rust. A possible reason for this observation is that the two types of intracellular structures arose independently in some macrocyclic rust ancestor instead of one evolving directly from the other. However, it should be pointed out that only a limited number of rusts have been examined, either with the light or electron microscope, and there may be more variation even in D-haustoria than now imagined. Some indication of this is provided by the light microscopic observation that urediospore-derived infections of *Ravenelia humphreyana* give rise to a primarily intracellular mycelium (Hunt, 1968). How this mycelium relates structurally, functionally, and phylogenetically to the typical D-haustorium is unknown. A phylogenetic series, ranging from a haustorium with neck and spherical body, through the more filamentous intracellular structures, to a completely intracellular mycelium has been suggested to exist among dikaryotic, tropical, rusts (Rajendren, 1972). However, since some of the species examined were microcyclic and of uncertain sporic origin, part of this range of morphologies may reflect the difference between the intracellular structures of basidiospore-derived infections and those in other stages of the life cycle. In addition this phylogenetic scheme conflicts with the pronounced ultrastructural specialization of what Rajendren suggests to be the most primitive type of haustorium. Only more detailed investigations, involving both light and electron microscopy, of a wide variety of rust genera and species at known stages in their life cycle will provide a true picture of the variation in intracellular structures in this group of fungi.

In spite of the differences between M- and D-haustoria, the ultrastructural host responses to these two structures is remarkably similar in many ways, particularly in terms of features associated with the host–parasite interface. One of the few obvious and reasonably consistent differences seems to be the greater degree of synthesis of wall-like materials around the M-haustorium (either in the form of fibrillar Type II collars, or the greater amount, and earlier production, of electron-opaque materials in the extrahaustorial matrix). If one assumes this synthesis to be a sign of greater incompatibility (see Chapter 6, Section I,B,1), this too may reflect the lack of specialization of M-haustoria to the intracellular situation. It is of interest that in this, and other, ultrastructural features, the M-haustorium is remarkably similar to the intracellular structures of the smut fungi (Fullerton, 1970) and the vesicular–arbuscular mycorrhizae (e.g., Cox and Sanders, 1974; Kinden and Brown, 1975b), both in fungal structure and host response; possibly a similar type of host–parasite interaction may exist in each of these situations.

Whatever the nature of the interactions between rusts and their host, the interface between the two organisms is of obvious importance in governing these processes. From a survey of ultrastructural studies, Bracker and Littlefield (1973) have recognized 39 types of interfaces which may exist between eukaryotic symbionts and their hosts and at least 12 of these occur during the formation of the D-haustorium from the haustorial mother cell. There is ultrastructural evidence that the host–parasite interaction may not be the same at each interface (e.g., the host appears to respond differently to the neck and body of D-haustoria, see Section II,C,1, this chapter), and different responses at different interfaces may explain the variation in ultrastructural details seen between various incompatible associations (see Chapter 6). Logically, it is the interactions which take place across some of these interfaces which determine the success of the infection (some of the more important interfaces in this respect may be discerned from the suggested critical stages of infection shown in Chapter 6, Fig. 1, Section I,C); however, information concerning the nature of such localized interactions is difficult to obtain by means other than electron microscopy. Freeze-etching and the use of the plasmalemma (PACP) stain are two examples of techniques which have already provided information difficult to obtain by standard biochemical procedures and it is expected that the application, or development, or other ultrastructural techniques will prove particularly fruitful in the future for revealing hitherto unknown features of the interactions between the rusts and their hosts.

5

Nuclei

Diagrams for the isometric projection of chromosome distribution and changes in morphology of the nucleus associated organelle are shown in Figs. 1 and 2. These topics will be discussed further in Section II.

I. INTERPHASE NUCLEI

A. Chromatin

The ultrastructure of interphase nuclei of the rusts closely resembles that of other eukaryotic organisms. Each nucleus consists of a lightly-staining matrix surrounded by a double membrane containing nuclear pores. Small patches of electron-opaque material, presumed to be condensed chromatin, seem to be characteristically present in the nucleoplasm of parasitic hyphae (Fig. 3) (Coffey, 1975; M. C. Heath and I. B. Heath, 1978; Mendgen, 1973a; Rijo and Sargent, 1974; Walles, 1974) but a detailed study of nuclei of *Uromyces phaseoli* var. *vignae* (M. C. Heath and I. B. Heath, 1978) has shown striking differences in the degree of condensation (under the same fixation conditions) between different stages in fungal development (compare Figs. 3 and 4). Unfortunately, few other workers have specifically examined this phenomenon and any conclusions drawn from a comparison of published micrographs must be viewed with caution due to the variety of fixation procedures used to prepare the material. Nevertheless, such a comparison does show that *Puccinia graminis* f. sp. *tritici* (Manocha and Wisdom, 1971) resembles *U. phaseoli* var. *vignae* in the marked dispersal of chromatin during germ tube growth from urediospores, while the striking increase in condensation seen in young haustorial mother cells of the latter fungus (uredial stage) can also be seen in nearly all published studies where this structure is illustrated (*Melampsora lini*, *Puccinia helianthi*, Coffey *et al.*, 1972a;

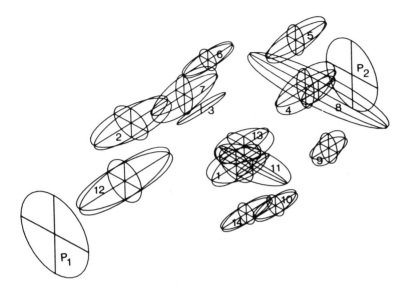

Fig. 1. Isometric projection (made by a computer driven plotter) of the distribution of chromosomes in a serially-sectioned, metaphase spindle of *Uromyces phaseoli* var. *vignae*, part of which is shown in the micrograph, Fig. 21. The polar, disk-shaped nucleus associated organelles are designated P_1 and P_2. The close clustering of some of the supposed chromosomes is, in part, an artifact of the projection. (From Heath and Heath, 1976.)

Hemileia vastatrix, Rijkenberg and Truter, 1973a; *U. phaseoli* var. *typica,* Mendgen, 1975). In *U. phaseoli* var. *vignae,* the degree of condensation appears more variable once the haustorial mother cell nuclei enter the haustorium (M. C. Heath and I. B. Heath, 1978) and, again, this may be the case for other rusts (compare micrographs of *U. phaseoli* var. *typica* of Hardwick *et al.,* 1971, with those of Mendgen, 1975, and micrographs of *M. lini* of Coffey, 1975, with those of Coffey *et al.,* 1972a). In studies of sporulation, chromatin has been reported to disappear immediately after nuclear fusion in the developing teliospore of *Gymnosporangium juniperi-virginianae* (Fig. 5) (Mims, 1977b) but appears highly condensed in the mature teliospore (Fig. 6) (Mims *et al.,* 1975). Highly condensed chromatin has also been reported in pycniospores of *G. juniperi-virginianae* (Mims *et al.,* 1976) and *Puccinia sorghi* (Rijkenberg and Truter, 1974a) but aeciospores (Rijkenberg and Truter, 1974b) and urediospores (Van Dyke and Hooker, 1969) of this latter fungus and urediospores of *U. phaseoli* var. *vignae* (M. C. Heath and I. B. Heath, 1978), *P. coronata* f. sp. *avenae* and *P. graminis* f. sp. *avenae* (Harder, 1976c) do not show this feature.

Taken as a whole these observations suggest that there may be some tem-

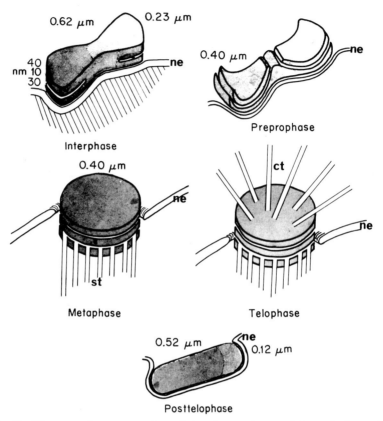

Fig. 2. Diagrammatic representation of the changes in morphology of the nucleus associated organelle (NAO) through the mitotic cycle of *Uromyces phaseoli* var. *vignae*. The cross-hatching under the interphase NAO indicates the differentiated region of nucleoplasm typically associated with interphase NAOs. ne, Nuclear envelope; st, spindle microtubules; ct, cytoplasmic microtubules. (From Heath and Heath, 1976.)

poral consistency between species in these changes in the appearance of the chromatin. Therefore, further investigation may be worthwhile particularly since these changes must reflect some physiological change within the nucleus. However, with the current lack of correlative physiological information, it is difficult to assess what these changes may be since it is conceivable that one is not observing differences in chromatin organization but rather changes in the response of the nucleoplasm as a whole to the preparative procedures employed (M. C. Heath and I. B. Heath, 1978). It is also possible that some (but not all) stages showing condensed chromatin merely reflect the smaller volume of the nucleus (see below) and therefore closer packing of its contents. Interpretations are further complicated by the possibility that some of this "condensed chromatin" may not be chromatin at all, since electron-opaque material has been observed in the mitotic nucleus of *Uromyces phaseoli* var. *vignae* which seems distinct from the chromosomes and is not associated with either the spindle or the nucleolus (Heath and Heath, 1976). Even assuming that most of the condensed chromatin is what its name implies, it remains to be determined whether such condensation indicates a lack of transcription as suggested for some animal cells (Frenster, 1974). Certainly the converse association of dispersed chromatin with high transcriptional activity (Frenster, 1974) is questionable, at least in *U. phaseoli* var. *vignae,* since such dispersion in the nuclei of the germ tube coincides with what has been shown in the closely related *U. phaseoli* var. *typica* to be a period of relatively low nucleic acid synthesis (see references in M. C. Heath and I. B. Heath, 1978).

B. Nucleoli

In all the rusts so far examined, nucleoli are commonly, if not always, present in nuclei of parasitic hyphae and haustoria. Like those of other organisms, these nucleoli contain electron-opaque fibrillar and granular components, often interspersed with electron-lucent lacunae (Fig. 7) (Harder, 1976a; M. C. Heath and I. B. Heath, 1978; Wright *et al.,* 1978). Very little is known, however, about the detailed ultrastructure of nucleoli in other stages of the rust life cycle except the urediospore and the subsequently formed germ tube and infection structures. In the urediospore, the presence or absence of the nucleolus has been the subject of controversy for some years. Both light and electron microscopy have shown nucleoli in developing urediospores (*Melampsora lini,* Littlefield, unpublished; *Puccinia graminis* f. sp. *tritici;* Mitchell and Shaw, 1969; Thomas and Isaac, 1967; *P. helianthi,* Craigie, 1959; Harder, 1976c) but there are conflicting ultrastructural reports of their presence in mature urediospores and subsequent stages of fungal growth prior to entry into the host plant (e.g., the differing reports for *P. graminis* f. sp. *tritici* by Manocha and Wisdom, 1971 and Dunkle *et al.,* 1970). Their supposed absence before the establishment of an association with the host has led to the hypothesis that the latter in some way stimulates nucleoli formation (Manocha and Wisdom, 1971; Shaw, 1967). Whether there is indeed some stage of urediospore maturation during which the nucleoli disperse remains to be determined, but more recent, detailed studies have shown that, at least for *Uromyces phaseoli* var. *vignae* (M. C. Heath and I. B. Heath, 1978), *U. phaseoli* var. *typica* (Mendgen, personal communication), and *P. graminis* f. sp. *tritici* (Robb, personal communication), nucleoli are present at all stages of early growth from the urediospore. In both *U. phaseoli* var. *vignae* and *P. graminis* f. sp. *tritici,* nucleoli are ini-

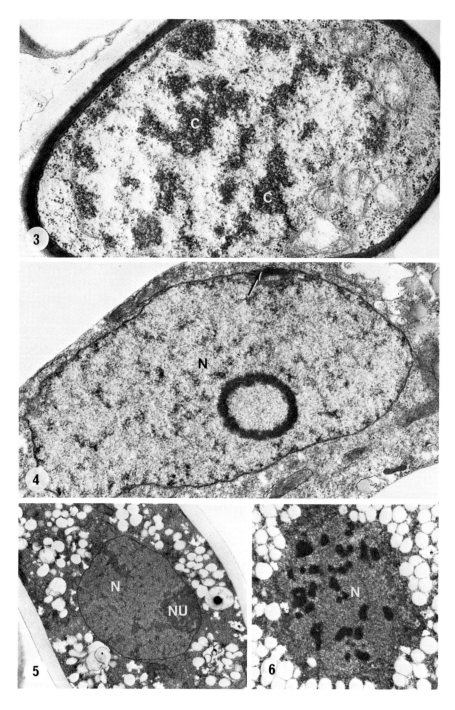

tially fibrillar (and often vacuolate in *U. phaseoli* var. *vignae)* (Figs. 4 and 8) but become granular (Fig. 9) during the formation of the substomatal vesicle (*P. graminis* var. *tritici,* Robb, personal communication) or the first haustorial mother cell (*U. phaseoli* var. *vignae,* M. C. Heath and I. B. Heath, 1978) (see Chapter 3, Section II for the developmental relationship between these structures).

Nucleolar granules are generally considered to be ribosome precursors, and predominantly fibrillar nucleoli are commonly associated with periods of relative inactivity in ribosome synthesis (Smetana and Busch, 1974). Thus the lack of granules during urediospore germination and the early stages of infection structure formation in *Uromyces phaseoli* var. *vignae* correlates well with the biochemical data from *U. phaseoli* var. *typica* which suggests that few ribosomes are synthesized during this period (Yaniv and Staples, 1969). It is significant that both the Heath and Robb studies mentioned above involved infection structures induced to form away from the host plant (see Chapter 3, Sections II and III); thus the observed eventual appearance of nucleolar granules indicates that the initiation of nucleolar activity depends more on the stage of growth of the fungus than on any stimulation from the host plant. This suggestion is further supported by the observation that when *Melampsora lini* is grown in axenic culture, the originally small, fibrillar nucleoli increase in size and granularity (and presumably activity) as the culture becomes established (Manocha, 1971). The molecular basis for axenic growth of certain rusts is as yet unknown and it is therefore intriguing that Harder (1976a) claims that nucleoli of *Puccinia graminis* f. sp. *tritici* and *P. coronata* show less ultrastructural zonation in axenic culture than in

---

Figs. 3–6. Changes in the appearance of chromatin during different stages of the rust life cycle.

Fig. 3. A nucleus in an intracellular hypha of *Uromyces phaseoli* var. *vignae* (uredial stage). Electron-opaque patches (C), presumed to be condensed chromatin, are scattered throughout the nucleoplasm. (× 27,900) (M. C. Heath, unpublished.)

Fig. 4. A nucleus (N) in a young germ tube of *U. phaseoli* var. *vignae*. Note the dispersed appearance of the chromatin compared with Fig. 3. The nucleus associated organelle (arrow) and a spherical, vacuolate nucleolus (see Fig. 8) can also be seen in this micrograph. (× 14,500) (From M. C. Heath and I. B. Heath, 1978. Reproduced by permission of the National Research Council of Canada from *Can. J. Bot.* **56**, 648–661.)

Fig. 5. Young teliospore of *Gymnosporangium juniperi-virginianae* after nuclear fusion. Note the dispersed appearance of the chromatin in the nucleus (N) and the nucleolus (NU) appressed to the nuclear envelope. (× 11,000) (From Mims, 1977b. Reproduced by permission of the National Research Council of Canada from *Can. J. Bot.* **55**, 2319–2329.)

Fig. 6. Older teliospore of *G. juniperi-virginianae* than that shown in Fig. 5. Note the condensed appearance of the chromatin in the nucleus (N). (× 13,000) (From Mims *et al.*, 1975. Reproduced by permission of the National Research Council of Canada from *Can. J. Bot.* **53**, 544–552.)

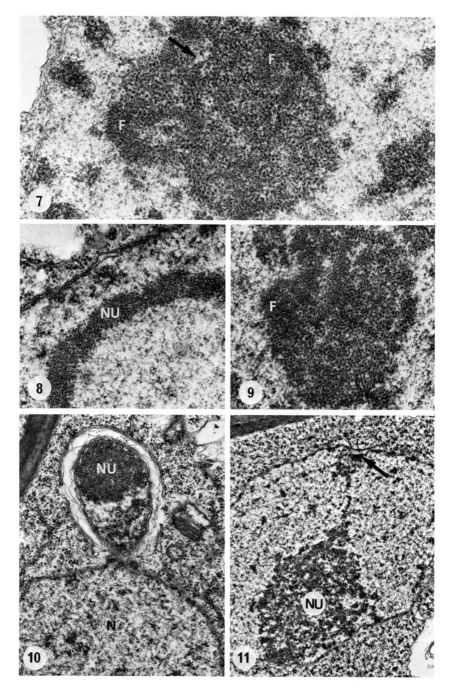

parasitic phases; he compares the appearance of the nucleoli in axenic culture to those seen in some animal tumor cells. The generality of this observation, and its significance, awaits investigation.

Although not widely reported for other rust genera, nuclear profiles suggesting the complete or partial extrusion of the nucleolus via a lobe of the nucleus, are apparently common in parasitic and axenically grown hyphae of several *Puccinia* spp. (Fig. 10) (Harder, 1976a; Wright *et al.*, 1978). This phenomenon may partly explain the pronounced differences in nuclear size frequently noted by light microscopists (e.g. the "expanded" and "unexpanded" nuclei reported by Savile, 1939) although some of the smaller nuclei may simply be ones which have not undergone the usual increase in volume following nuclear division (M. C. Heath and I. B. Heath, 1978). The relationship between nucleolar extrusion and nuclear division is discussed later.

C. Nucleus-associated Organelle

Studies of a number of rusts from several genera have shown that all have a similar and characteristic nucleus associated organelle* (NAO) situated in a slight depression of the nuclear envelope (Dunkle *et al.*, 1970; Coffey *et al.*, 1972a; Harder, 1976a; Heath and Heath, 1976; Mims, 1977b; Wright *et al.*,

---

*This structure has had numerous names (e.g., centrosome, microtubule organizing center, kinetochore equivalent) since its discovery in the fungi. We use the term NAO (Girbardt and Hädrich, 1975), rather than the recently popular spindle pole body, since it is functionally neutral and locationally more correct considering the structure spends most of its time *not* associated with the spindle.

Fig. 7. Typical nucleolus from an intercellular hypha of *Uromyces phaseoli* var. *vignae*. Note the fibrillar areas (F) interspersed with granular material and also the more translucent lacunae (arrow) which permeate the nucleolus. ($\times$ 40,500) (M. C. Heath, unpublished.)

Fig. 8. The same vacuolate, germ tube nucleolus (NU) as shown in Fig. 4. Note the almost complete absence of nucleolar granules. ($\times$ 45,600) (From M. C. Heath, unpublished.)

Fig. 9. A nucleolus from an infection hypha which has also produced a haustorial mother cell and secondary hypha on an artificial membrane. The nucleolus is now nonvacuolate and contains both fibrillar (F) and granular components. ($\times$ 35,400) (M. C. Heath and I. B. Heath, 1978. Reproduced by permission of the National Research Council of Canada from *Can. J. Bot.* 56, 648–661.)

Fig. 10. Profile suggesting a stage in the extrusion of the nucleolus (NU) from the nucleus (N) of an intracellular hypha of *Puccinia coronata*. ($\times$ 26,000) (From Harder, 1976a. Reproduced by permission of the National Research Council of Canada from *Can. J. Bot.* 54, 981–994.)

Fig. 11. A developing urediospore of *P. coronata* showing a threadlike connection between the nucleolus (NU) and the region of the nucleus close to the nucleus associated organelle (arrow). ($\times$ 18,400) (From Harder, 1976a. Reproduced by permission of the National Research Council of Canada from *Can. J. Bot.* 54, 981–994.)

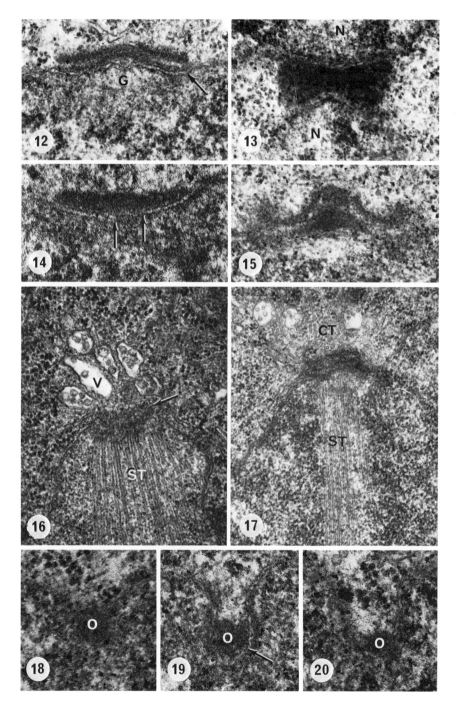

1978). Where examined in detail, this NAO consists of two, somewhat acir-
cular, and often apparently three-layered discs connected by an electron-
opaque bar (Figs. 12-14) (Harder, 1976a; Heath and Heath, 1976).
Although technically bipolar, this structure is fairly distinct from the
diglobular form characteristic of many higher basidiomycetes (see review by
Heath, 1978). It also differs from those in some other basidiomycetes (e.g.,
Girbardt, 1968; McCully and Robinow, 1972a,b; Poon and Day, 1976) in
having few or no cytoplasmic microtubules associated with it during inter-
phase (Coffey et al., 1972a; Harder, 1976a; Heath and Heath, 1976). The
NAO is associated, however, with a hemisphere of granular material on the
inside of the nucleus (Fig. 12) (Dunkle et al., 1970; Coffey et al., 1972a;
Harder, 1976a; Heath and Heath, 1976; Wright et al., 1978) and for some
Puccinia spp., Harder (1976a) and Wright et al. (1978) claim a direct connection between the two structures through a gap in the nuclear envelope
(Fig. 14), although no such gap has been seen in U. phaseoli var. vignae (Fig.
12) (Heath and Heath, 1976). This material on the inside of the nucleus
usually stains more lightly than other electron-opaque components of the

---

Figs. 12-20.   The nucleus associated organelle (NAO). For diagrammatic representation of
changes in NAO morphology associated with mitosis, see Fig. 2.
Fig. 12.   Longitudinal section of a NAO adjacent to an interphase nucleus of Uromyces
phaseoli var. vignae. Note the hemisphere of granular (G) material just inside the nucleus and
the more electron-opaque material where this material borders the nucleoplasm. The nuclear
envelope (arrow) appears intact where it underlies the NAO. (× 52,600) (From Heath and
Heath, 1976.)
Fig. 13.   A surface view of the interphase NAO of U. phaseoli var. vignae. Since the NAO is
situated in an indentation of the nucleus, the latter appears in two portions (N) in this
micrograph. (× 58,400) (From Heath and Heath, 1976.)
Fig. 14.   The interphase NAO of Puccinia coronata with what appears to be a pore (arrows) in the underlying nuclear envelope (compare with Fig. 12). (× 72,000) (From Harder,
1976a. Reproduced by permission of the National Research Council of Canada from Can. J.
Bot. 54, 981-994.)
Fig. 15.   A preprophase NAO of U. phaseoli var. vignae. The terminal disks of the NAO
(see Fig. 13) seem to be separating and serial sections show that each has increased in diameter
(this section is not quite longitudinal). (× 58,400) (From Heath and Heath, 1976.)
Fig. 16.   The pole of a longitudinally sectioned, metaphase spindle of U. phaseoli var.
vignae showing spindle microtubules (ST), the now disk-shaped NAO (arrow) situated within a
gap in the nuclear envelope, and numerous multivesicular bodies (V) in the cytoplasm. Note the
absence of cytoplasmic microtubules in the vicinity of the NAO (compare with Fig. 17).
(× 49,800) (From Heath and Heath, 1976.)
Fig. 17.   The pole of a longitudinally sectioned, telophase spindle of U. phaseoli var. vignae
showing spindle microtubules (ST), and numerous cytoplasmic microtubules (CT) radiating
from the NAO into the cytoplasm. (× 35,600) (From Heath and Heath, 1976.)
Figs. 18-20.   Transverse sections from the beginning, middle, and end respectively of serial
sections cut through a recently posttelophase NAO (O) of U. phaseoli var. vignae. Note the deep
indentation of the nuclear envelope (arrow). (All × 82,800) (From Heath and Heath, 1976.)

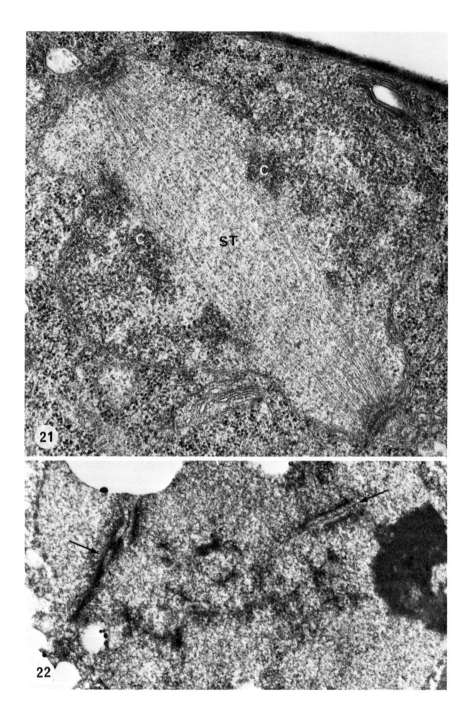

nucleoplasm but is commonly associated with more densely-staining material on the surface away from the nuclear envelope (Fig. 12) (Coffey *et al.*, 1972a; Harder, 1976a; Heath and Heath, 1976; Wright *et al.*, 1978). The latter material is reported to have loose connections to the nucleolus in several *Puccinia* spp. (Fig. 11) (Harder, 1976a) and this observation may be of relevance to the claim by Girbardt (1968, 1971) that nucleoli of some living fungi behave as if attached in some way to the NAO.

II. NUCLEAR DIVISION

To date, only three detailed ultrastructural studies have been carried out on nuclear division in the Uredinales and all examined mitosis in dikaryotic hyphae (Harder, 1976b; Heath and Heath, 1976; Wright *et al.*, 1978). From nuclear configurations suggesting the extrusion of nucleoli from nonmitotic nuclei (Fig. 10), Harder (1976a) and Wright *et al.* (1978) conclude that for *Puccinia* spp. this phenomenon is a prelude to mitosis as has been suggested to be the case for some higher basidiomycetes (Girbardt, 1968). However, none of these authors provide evidence that these nuclei were about to divide. In infection structures of *Uromyces phaseoli* var. *vignae* where mitosis occurs at certain stages of development and is therefore highly predictable, correlated light microscopy of living material and electron microscopy suggest that nucleolar extrusion begins during anaphase (M. C. Heath and I. B. Heath, 1978), although it is uncertain whether actual separation occurs then or is delayed until telophase (Heath and Heath, 1976). Interestingly, the few other reports of nucleolar expulsion *during* division in the fungi have been for ascomycetes (e.g., Hung and Wells, 1977).

The first sign of mitosis in *Uromyces phaseoli* var. *vignae* is the separation of the two discs of the NAO (Fig. 15), each of which enlarges and becomes positioned at the poles of the metaphase spindle, lying *within* a gap in the otherwise intact nuclear envelope (Fig. 16) (Heath and Heath, 1976). Stages prior to metaphase were not found for the *Puccinia* spp. (Harder, 1976b; Wright *et al.*, 1978), but during metaphase in these and *U. phaseoli* var. *vignae* (Heath and Heath, 1976), multivesicular bodies aggregate in the

---

Fig. 21. One of the series of sections of a metaphase spindle of *Uromyces phaseoli* var. *vignae* from which Fig. 1, Chapter 5 was prepared. Note the central core of the spindle microtubules (ST) and the peripheral arrangement of the chromosomes (C). Serial sections showed that this spindle was occupying only a small portion of the nucleus; the rest occurs out of the plane of this section. (× 39,500) (M. C. Heath and I. B. Heath, unpublished.)

Fig. 22. Synaptonemal complexes (arrows) in a diploid nucleus of a teliospore of *Gymnosporangium juniperi-virginianae*. This spore was still within the telium. (× 25,000) (From Mims, 1977b. Reproduced by permission of the National Research Council of Canada from *Can. J. Bot.* 55, 2319-2329.)

cytoplasm near the NAOs (Fig. 16). The function of these bodies is unknown but their presence in these two different genera, as well as in some other basidiomycetes during mitosis (McCully and Robinow, 1972b) may indicate a significant role during mitosis.

The details of the metaphase spindle (Fig. 21) have been analyzed in detail only in *Uromyces phaseoli* var. *vignae* (Heath and Heath, 1976). Pole-to-pole, interdigitating, pole-to-chromosome, and fragmentary microtubules are present and arranged in a central bundle along the surface of which lie the chromosomes. The latter appear to be attached by up to three microtubules per kinetochore. As is common in most groups of fungi (Heath, 1978), the metaphase chromosomes (i.e., those with microtubule connections to both poles) are not aggregated to form a metaphase plate. A computer-plotted, three-dimensional reconstruction of the chromatin, derived from the analysis of serial sections from one of these metaphase spindles, shows 14 individual clusters (Fig. 1 on page 195) suggesting that this is the haploid number for this species. This figure is much higher than that based on light microscopy for other rusts where 6 is the commonly reported number (e.g., Valkoǔn and Bartoš, 1974).

At telophase, the pole-to-pole microtubules of *U. phaseoli* var. *vignae* elongate and the nucleoplasm (what is left after nucleolar extrusion) becomes aggregrated into two masses, each at the end of the long spindle. At this time the polar NAOs become associated with about 60 cytoplasmic microtubules running predominantly away from the nucleus (Fig. 17). Similar associations in the fungi have been suggested to aid telophase nuclear elongation through some type of microtubule-cytoplasm interaction (Heath, 1978) but such microtubules were not seen in the otherwise similar telophase stage in the *Puccinia* spp. (Harder, 1976b).

In all the rusts examined in the Harder (1976b) and Heath and Heath (1976) studies of mitosis, the ultrastructural evidence suggests that the two daughter nuclei form by constriction of the nuclear envelope around the chromatin which by now is amassed at the spindle poles; the remaining central portion of the spindle presumably disintegrates. Following telophase, the NAO in *Uromyces phaseoli* var. *vignae* loses its associated cytoplasmic microtubules and becomes completely amorphous and sausage-shaped (Figs. 18–20) before resuming its typical interphase appearance; the comparable behavior of the *Puccinia* NAOs has not been reported but a similar phenomenon occurs in the basidiomycete, *Polystictus versicolor* (Girbardt and Hadrich, 1975). A diagram summarizing the changes in morphology of the NAO during mitosis in *U. phaseoli* var. *vignae* is shown in Fig. 2 (page 196).

The ultrastructural events associated with nuclear fusion in the teliospore and the subsequent meiotic nuclear division are virtually unknown. The only ultrastructural report on meiosis in the rusts to date is that of Mims (1977b) who has shown synaptonemal complexes in teliospores of *Gymnosporangium juniperi-virginianae* (Fig. 22). It is interesting that these were observed soon after nuclear fusion while nuclear division is thought to take place much later when the teliospore germinates. Such interruptions in meiosis are common in animal oocytes (e.g., Smith, 1975) but have not been previously documented for any true fungus.

In conclusion, nuclear division (or at least mitosis) in the rusts ultrastructurally falls within the range of types found in other septate fungi (Heath, 1978). Certain aspects (the intact nuclear envelope, the compact core of microtubules and the morphology of the NAO at metaphase), however, more closely resemble mitosis in ascomycetes than basidiomycetes (Heath and Heath, 1976) and thus may provide additional evidence for the phylogenetic relationship of these two groups.

6

Incompatible Plant-Rust
Associations

I. "NATURAL" RESISTANCE

A. Fungal Behavior on the Plant Surface

Several reports from light microscope studies indicate that the behavior of urediospore germ tubes on the plant surface may play a role in the resistance of nonhost plant species (i.e., those for which the rust is not a pathogen) (Heath, 1974; Heath, 1977; Wynn, 1976). The scanning electron microscope (SEM) has proved particularly useful in examining this phenomenon and has shown that the lack of directional growth (see Chapter 3, Section I) by germ tubes of *Uromyces phaseoli* var. *typica* on wheat leaves correlates with the lack of adherence of these germ tubes to the leaf surface, and also with the presence of wax extrusions on the leaf surface (Wynn, 1976). In contrast, germ tubes of the wheat pathogen, *Puccinia graminis* f. sp. *tritici* do not show surface adhesion or directional growth on the host leaf unless this wax is present (see Chapter 3, Figs. 4 and 5) (Wynn and Wilmot, 1977). This obvious adaptation of a rust to the surface of its host is further exemplified by the ability of germ tubes of *U. phaseoli* var. *typica* to grow right over stomata on wheat or oat leaf replicas (to which the germ tubes *do* adhere) without "recognizing" the stoma and forming an appressorium (Fig. 3). The suggested reason for this is the poorer development of the stomatal lips in the gramineous nonhosts compared with those of the host plant (compare Figs. 2 and 3) (Wynn, 1976).

Figure 1 presents a proposed scheme of events during infection. It will be discussed in detail in Section I,C.

The surface behavior of germ tubes from rust spores other than urediospores has been seldom studied in incompatible associations. However, in an SEM investigation of "reduced needle lesion" type of resistance of *Pinus*

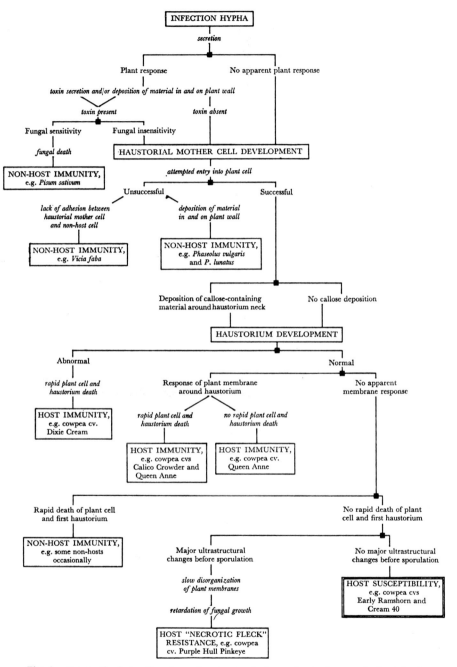

Fig. 1. Proposed scheme (based on ultrastructural observations) of events during infection by *Uromyces phaseoli* var. *vignae* which could lead to the observed resistance or susceptibility of host and nonhost plants. It is proposed that the responses induced at each "switching point" (■) determine the subsequent progress of infection. (From Heath, 1974a.)

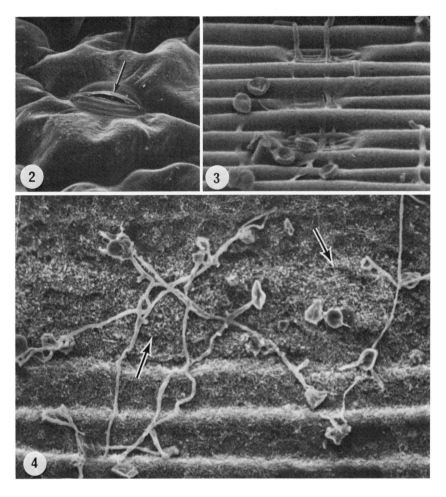

Figs. 2–4. Scanning electron micrographs of the surface behavior of germ tubes on resistant host or nonhost plants.

Fig. 2. Typical stoma of *Phaseolus vulgaris* showing the protruding stomatal lips (arrow). (× 580) (Fixed in glutaraldehyde and osmium tetroxide, courtesty of W. K. Wynn, unpublished.)

Fig. 3. Urediospore germ tubes of *Uromyces phaseoli* var. *typica* growing across the epidermal ridges of a replica of a nonhost *Triticum aestivum* leaf. Note the lack of appressoria over the stomata, and the fact that the latter seem to lack protruding lips characteristic of the host plant (see Fig. 2). (× 340) (From Wynn, 1976.)

Fig. 4. Basidiospore germ tubes of *Cronartium ribicola* on a leaf of *Pinus strobus* which exhibits "reduced needle lesion" resistance. The germ tubes have grown across the completely wax-occluded stomata (arrows) without attempting to enter the leaf. (× 660) (Freeze-dried, courtesy of R. F. Patton, unpublished.)

strobus to *Cronartium ribicola,* epicuticular wax, exuded from the subsidiary cells, was seen to partially or completely occlude the opening to the stoma; this apparently prevented ingress into the leaf by the basidiospore germ tube (Fig. 4) (Patton and Spear, 1977).

B. Interactions within the Tissue

In spite of the above-mentioned investigation of "reduced needle lesion" resistance, light microscope studies have generally suggested that rust development on the leaf surface is not a determinant in the resistance of host cultivars to either monokaryotic or dikaryotic phases (Hilu, 1965; Melander and Craigie, 1927; Rothman, 1960; Zimmer, 1965) and that at least some attempts at ingress into the tissue commonly occur even in nonhost plants (Heath, 1977). Light microscope and biochemical studies of the events which then take place suggest that this type of resistance is physiologically, not physically, based. However, it is probably fair to say that neither the reactions which directly lead to the limitation of fungal growth, nor the basis of host-parasite specificity, are clearly understood. The few transmission electron microscope studies of this "physiological" resistance have also proved to be rather disappointing in providing unequivocal answers to these problems, probably because of the same difficulty in distinguishing cause from effect which plagues other types of investigations (for a recent discussion exemplifying this point see Heath, 1976a) and also because physiological changes are not always reflected in changes in ultrastructure. Nevertheless, the electron microscope has provided information unattainable by other means and at least has added a structural dimension to the concept of the ways in which a resistant plant may respond to rust attack.

1. Cultivar Resistance

One of the few generalizations which can be made from light microscope studies of resistance is that invasion of resistant cultivars of the host species usually results in the fungus producing at least one haustorium (R. F. Allen, 1923b; Hilu, 1965; Jones and Deverall, 1977; Littlefield, 1973; Mendgen, 1978; Zimmer, 1965) while haustoria are rare or absent in nonhost species (Gibson, 1904; Heath, 1972, 1974a, 1977; Leath and Rowell, 1966, 1969). If haustoria are formed, resistance is usually, but not always (Heath, 1971), manifested by the death of the invaded cell (see above references) and sometimes the subsequent death of adjacent noninvaded cells (Heath, 1972; Hilu, 1965; Littlefield and Aronson, 1969; Zimmer, 1965). However, particularly in "cultivar resistance" (i.e., resistance manifested by cultivars of the host species), the degree of fungal growth, the number of haustoria, and the extent of host necrosis can all vary considerably, even within the range of responses shown by different cultivar and strain combinations of single species of host and rust (Littlefield, 1973; Mendgen, 1978). Unfortunately, the majority of ultrastructural studies of cultivar resistance have focused on the type where the fungus shows considerable growth and where host necrosis is fairly extensive, commonly resulting in a macroscopic, easily detectable fleck on the infected tissue. To date there have been nine such studies of this intermediate type of resistance, all but one involving the dikaryotic stage of the fungus (*Melampsora lini,* Coffey, 1976; *Puccinia graminis* f. sp. *tritici,* Ehrlich and Ehrlich, 1962; *P. sorghi,* Van Dyke and Hooker, 1969; Manocha, 1975; Shaw and Manocha, 1965b; Skipp *et al.,* 1974; *P. carthami,* pycnial stage, Zimmer, 1970; *P. coronata* f. sp. *avenae,* Onoe *et al.,* 1976; *Uromyces phaseoli* var. *vignae,* Heath, 1972; *U. phaseoli* var. *typica,* Mendgen, 1977).

In all these studies there is either no initial detectable difference in the appearance of the haustorium, or in the appearance of the invaded cell, from the compatible interaction (Fig. 5) (Heath, 1972; Mendgen, 1977; Shaw and Manocha, 1965b) or the authors claim only rather subtle differences such as the swelling of haustorial mitochondria, an absence of fungal lomosomes, extrahaustorial membrane breakdown, and host vacuolation (Van Dyke and Hooker, 1969); reduction in host mitochondrial size, cristae disorganization, and alterations in chloroplast structure (Skipp *et al.,* 1974); fewer host mitochondria and less well developed endoplasmic reticulum (Ehrlich and Ehrlich, 1962); less haustorial glycogen and the disappearance of host crystal-containing microbodies (Zimmer, 1970); increased frequency of host Golgi bodies (Onoe *et al.,* 1976); and a more rapid loss of starch grains and the appearance of glycogenlike granules in the chloroplasts (Fig. 6) (Coffey, 1976). The validity and significance of these mainly quantitative differences between incompatible and compatible infections are difficult to assess since none of the ultrastructural studies except Onoe *et al.* (1976) give any numerical data and the sample size in some is seemingly rather small. Taken at face value, however, many of these observations could merely reflect the onset of necrosis of both the invaded cell and the haustorium and therefore cannot be used as unequivocal indicators of the primary responses determining resistance. The one possible exception to this is the claimed disappearance of the crystal-containing microbodies in resistant safflower seedlings (Zimmer, 1970). Since more of these microbodies occur in the resistant cultivar than in the susceptible, Zimmer suggests that their breakdown in the incompatible interaction releases autolytic enzymes which lead to the destruction of the cytoplasm of both the haustorium and invaded cell. Thus the fungus is restricted from colonizing the perivascular region, the main feature of this type of seedling resistance.

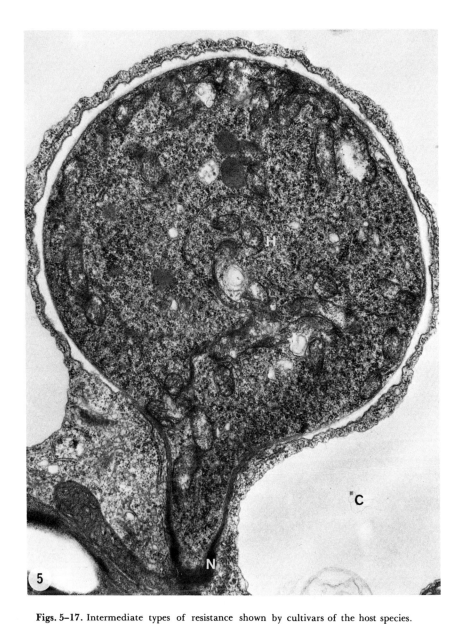

Figs. 5–17. Intermediate types of resistance shown by cultivars of the host species.

Fig. 5. The first formed D-haustorium of *Uromyces phaseoli* var. *typica,* 20 hours after inoculation of a *Phaseolus vulgaris* cultivar. This cultivar will eventually show a necrotic fleck, intermediate type of resistance. Structural features of both haustorium and host cell (C) resemble those in compatible associations. Note the presence of the neckband (N) (the proximal part of the neck is not shown in this section) and the way in which the haustorial wall becomes less electron-opaque, and less discrete, near the junction of the neck and body (see Haustorium Development, Chapter 4, Section II,A,2). (× 21,700) (M. C. Heath, unpublished.)

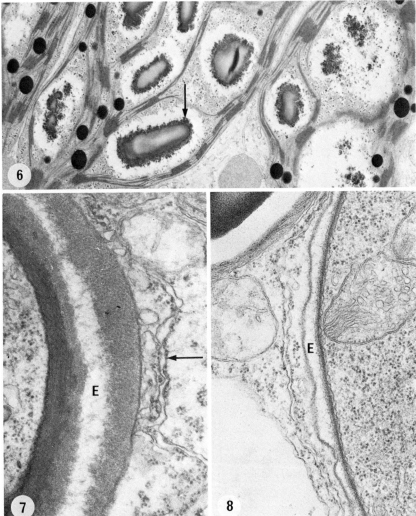

Fig. 6.   Glycogenlike granules (arrow) around starch grains of *Linum usitatissimum*. This cultivar exhibits intermediate resistance to *Melampsora lini*, conditioned by the K gene. (× 14,000) (From Coffey, 1976. Reproduced by permission of the National Research Council of Canada from *Can. J. Bot.* **54,** 1443–1457.)

Fig. 7.   The extrahaustorial matrix (E) in the same rust-host combination shown in Fig. 6. Note the dense, fibrillar material in the matrix which appears to be organized into two layers. Host endoplasmic reticulum (arrow) can be seen adjacent to the extrahaustorial matrix. (× 37,000) (From Coffey, 1976. Reproduced by permission of the National Research Council of Canada from *Can. J. Bot.* **54,** 1443–1457.)

Fig. 8.   The extrahaustorial matrix (E) in a compatible *M. lini*-host combination for comparison with Fig. 7 (× 35,000) (Courtesy of M. D. Coffey, unpublished.)

In addition to the possible cytoplasmic differences described above, electron-opaque, granular, or fibrillar material has been reported to accumulate faster in the extrahaustorial matrix of some of these incompatible interactions than in compatible ones (Figs. 7 and 8) (*Melampsora lini,* Coffey, 1976; *Puccinia graminis* f. sp. *tritici,* Manocha, 1975; Shaw and Manocha 1965b; *Uromyces phaseoli* var. *vignae,* Heath, 1972). In resistance to *P. graminis* f. sp. *tritici* conditioned by the Sr6 gene, autoradiography suggests that the development of this material, in both resistant and susceptible plants, coincides with the reduction of [³H] leucine incorporation into the fungus, and more inexplicably, into host cells (Figs. 9–12) (Manocha, 1975). Whether this extrahaustorial material represents host wall material, as suggested for similar material surrounding haustoria in compatible interactions involving the mycoparasite, *Piptocephalis virginiana* (Manocha and Lee, 1972; Manocha and Letourneau, 1978), or the endophyte of *Alder* sp. root nodules (Lalonde and Knowles, 1975) remains to be established. It also remains to be proven that the presence of this material is *directly* related to the reduction in label uptake by the haustorium: the material which develops during the intermediate type of resistance to *U. phaseoli* var. *vignae* certainly seems to lack the coherence necessary for a barrier to solute transport and it more or less disperses when the extrahaustorial membrane eventually disintegrates (Fig. 15) (Heath, 1972). In addition, a similar reduction in label uptake has also been reported for *U. phaseoli* var. *typica* in an intermediate type of resistant interaction (Mendgen, 1977) but, in this case, no mention is made of any correlation between this phenomenon and differences in the extrahaustorial matrix. Mendgen points out that this apparent reduced level of nutrient uptake by the fungus may not be a general phenomenon for all types of rust resistance and he has since found examples of other incompatible bean rust interactions where such a reduction in label uptake could not be demonstrated (Mendgen, personal communication).

Encasement of haustoria by electron-lucent material resembling that encasing haustoria in some compatible (see Chapter 4, Fig. 80) and some more highly incompatible (see Fig. 26) interactions has been mentioned only for the intermediate type of resistance to *Melampsora lini* studied by Coffey (1976). In contrast to the former situations, not all encased *M. lini* haustoria showed signs of necrosis, but by 9 days after inoculation, 20% of observed haustoria were encased.

In all of the above-mentioned incompatible interactions, the contents of haustorium-containing cells eventually disorganize. In the intermediate type of resistance to *Puccinia graminis* f. sp. *tritici,* Skipp *et al.* (1974) could only find completely disorganized haustoria within equally disorganized host cells (Fig. 16). They suggest that for this interaction, the death of the invaded cell must be rapid and must quickly bring about the death of the haustorium.

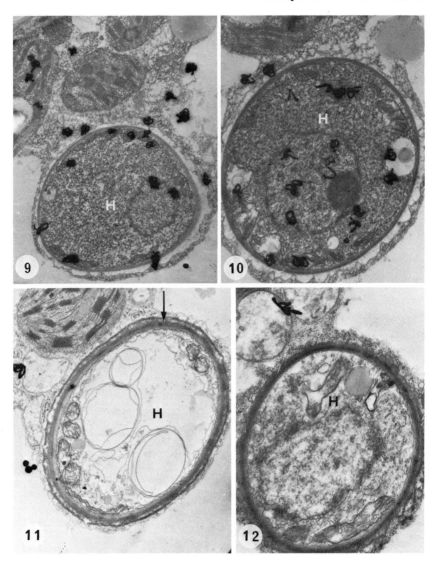

Figs. 9–11.   Haustoria (H) of *Puccinia graminis* f. sp. *tritici* in a susceptible host labeled with [³H] leucine at 6, 8, and 12 days after inoculation, respectively. The lack of silver grains in Fig. 11 correlates with an increase in the amount of staining material in the extrahaustorial matrix (arrow). (Fig. 9, × 9,000; Fig. 10, × 12,250; Fig. 11, × 11,250) (From Manocha, 1975.)

Fig. 12.   Haustorium of *P. graminis* f. sp. *tritici*; 4 days after inoculation of a [³H] leucine-labeled host showing an intermediate type of resistance. Note the absence of silver grains over the haustorium (H), and the presence of staining material in the extrahaustorial matrix. (× 14,800) (From Manocha, 1975.)

However, a different situation seems to exist in the superficially similar type of resistance to *P. carthami* (Zimmer, 1970) and *Uromyces phaseoli* var. *vignae* (Heath, 1972), where disorganization of the host cytoplasm appears more gradual and does *not* necessarily result in the immediate death of the haustorium (Fig. 13). In the latter study, host disorganization involves a dissection of the peripheral cytoplasm caused in part by the fusion of the endoplasmic reticulum and other membranes, with the plasmalemma (Fig. 14). The rarity of such fusion in normal plants or animals suggests that a major change in the properties of these membranes occurs in this host–parasite interaction. However, such a phenomenon has not been reported in any of the other types of intermediate resistance discussed above.

An interesting observation in light of the normal "modification" of the host plasmalemma around the haustorium (see Chapter 4, Section II,C,1) is the observation by Ehrlich and Ehrlich (1962) and Heath (1972) that this portion of the host plasmalemma is the last to disorganize. Again however, such an observation does not seem to be typical for the other studies of this type of resistance (Shaw and Manocha, 1965b; Van Dyke and Hooker, 1969).

Ultrastructural changes involving the formation of membranous whorls, a decrease in mitochondria size (Skipp *et al.*, 1974), or complete disorganization of the cytoplasm (Van Dyke and Hooke, 1969) have been reported in the host cells bordering those invaded cells also showing necrosis. In addition, Skipp *et al.* (1974) report the deposition of electron-opaque material between the plasmalemma of these cells and the cell wall (Fig. 17). This material resembles that observed in both resistant and susceptible reactions to *Puccinia graminis* f. sp. *tritici* by Ehrlich *et al.* (1968b) and also appears to be similar to the electron-opaque component of the callose-containing deposits and encasements found in invaded cells in the more extreme forms of resistance described below (Heath, 1971; Heath and Heath, 1971; Heath, 1972). Such deposits are most commonly interpreted as being a nonspecific response to injury (see Chapter 4, Section II,C, 2) which, when found in cells adjacent to necrotic areas, may help to limit the spread of damaging substances throughout the tissue (Skipp *et al.*, 1974).

In contrast to the relative popularity of the less severe types of cultivar resistance mentioned above, to our knowledge there have been only four ultrastructural studies of the more vigorous types which allow no more than one or two haustoria to develop and do not result in any macroscopic symptoms (El-Gewely *et al.*, 1972; Heath and Heath, 1971; Heath, 1972; Onoe *et al.*, 1976). All involve the dikaryotic stage in the life cycle and the most detailed studies (Heath and Heath, 1971; Heath, 1972) have focused on three different cowpea cultivars, two of which, when examined with the light microscope, seem to allow normal growth of the infection hypha but respond to the formation of the first haustorium by rapid necrosis of the invaded cell.

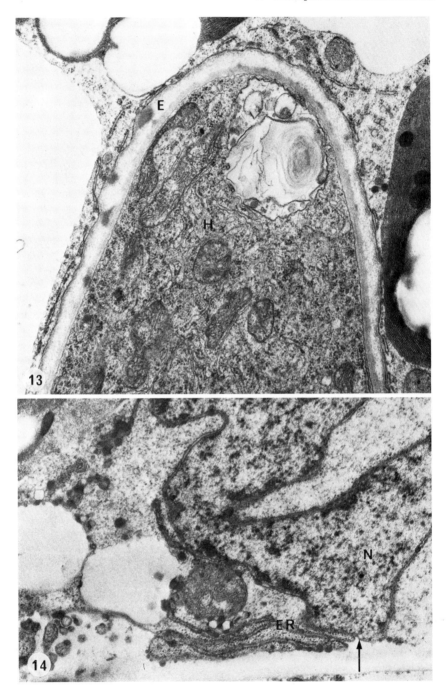

In some cases a secondary hypha is initiated and another attempt at haustorium formation is made, but this too elicits the same rapid host response and no further fungal growth occurs (Heath, 1971, 1972). In the third cultivar, Queen Anne, however, there is an interesting dimorphism in response to invasion in that the invaded cell only becomes necrotic in 60% of infection sites; in the remaining 40% the haustorium becomes completely enclosed in refractile, callose-containing material which builds up as a cup-shaped structure from the region of the host wall (Heath, 1971). However, in spite of the lack of host necrosis, this response seems to be equally effective in limiting fungal growth (Heath, 1971).

Electron microscopy shows that in all three cultivars, no host response can be detected before the formation of the first haustorium except for electron-lucent, callose-containing deposits in Queen Anne cells adjacent to some infection hyphae (Fig. 18) (Heath and Heath, 1971; Heath, 1972). In the two cultivars other than Queen Anne, and in infection sites of the latter showing the necrotic response, the high frequency of infection sites where the cytoplasm of both invaded cell and haustorium are completely disorganized (Fig. 19), even when the haustoria could not be more than 12 hours old, suggests that host necrosis is extremely rapid and involves the almost simultaneous death of the haustorium (Heath and Heath, 1971; Heath, 1972). As found for some types of intermediate resistance, cells adjacent to necrotic cells often also show signs of necrosis or at least deposits of electron-lucent material in the vicinity of plasmodesmata (Fig. 20) (Heath and Heath, 1971; Heath, 1972).

In spite of this general similarity between cultivars in response to infection, ultrastructural observations made on infection sites where the invaded cell had not yet disorganized suggests that there are, in fact, two distinct types of interaction (Heath, 1972). In one cultivar, haustorium development appears arrested before the neckband develops and the nuclei do not migrate from the haustorial mother cell (Fig. 22). However there is little initial sign of host reaction other than an aggregation of ribosome-studded, membrane-bound vesicles around the haustorial neck, small electron-opaque areas on most cell membranes, and a decrease in electron opacity of mitochondria matrices. In-

---

Fig. 13.   An early stage in the disorganization of a host cultivar showing intermediate resistance to *Uromyces phaseoli* var. *vignae.* The host cytoplasmic contents appear sparse but all membranes are still intact. Flocculent electron-opaque material can be seen in the extrahaustorial matrix (E) but the haustorial cytoplasm (H) resembles that in susceptible hosts. (× 26,400) (From Heath, 1972.)

Fig. 14.   A later stage in host cell disorganization than that shown in Fig. 13. Peripheral packets of cytoplasm are apparently being formed by the fusion of endoplasmic reticulum (ER) and the nuclear (N) envelope (arrow) with the plasmalemma. Note small drops of electron-opaque material on many of the cell membranes. (× 26,400) (From Heath, 1972.)

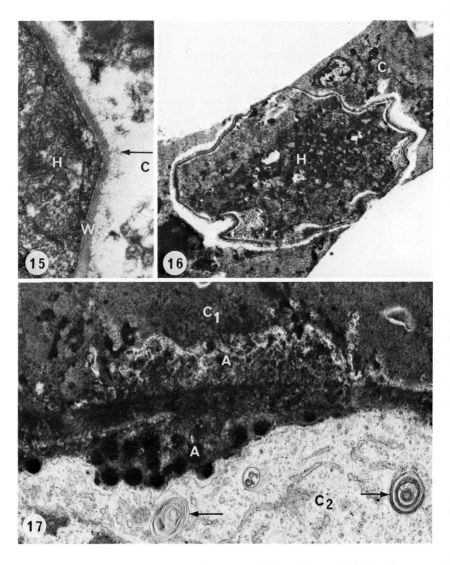

Fig. 15. A completely disorganized haustorium (H) and host cell (C) in the same host-parasite combination illustrated in Figs. 13 and 14. The extrahaustorial membrane has disappeared and the extrahaustorial matrix may possibly be discerned as diffuse material (arrow) next to the fungal wall (W). (× 38,000) (From Heath, 1972.)

Fig. 16. A completely disorganized haustorium (H) of *Puccinia graminis* f. sp. *tritici* in an equally disorganized host cell (C) 26 hours after inoculation of a host cultivar showing intermediate resistance conditioned by the Sr6 gene. The host cell contents have shrunken away from the cell walls (out of the field of view of this micrograph). (× 10,000) (From Skipp *et al.*, 1974. Reproduced by permission of the National Research Council of Canada from *Can. J. Bot.* 52, 2615-2620.)

deed, even those responses of aggregation of host cytoplasm and endoplasmic reticulum around the haustorium, normally found in the susceptible response, appear absent. In contrast, in the other two cultivars (including Queen Anne), full-sized, apparently normal haustoria are formed although those in Queen Anne lack neckbands (Fig. 32) [an interesting observation considering that these haustoria have otherwise developed beyond the stage where this structure should have formed (see Chapter 4, Section II,A,2) (Heath and Heath, 1971)]. Signs of host resistance are manifested soon after penetration, however, since electron-lucent material is deposited as a collar around the point of entry of the haustorium into the cell (Fig. 23). Like similar deposits formed in compatible interactions (see Chapter 4, Section II,C,2) this deposit usually contains membranous material but also often has a more electron-opaque, granular layer next to the host cell wall. The fact that the invaginated plasmalemma covers both the *inside* as well as the outside of this collar (i.e., it is a Type I collar, see Section II,C,2, Chapter 4 and it is *not* deposited directly against the fungal wall) suggests that this material represents a wound response to the hole in the host wall rather than a direct response to the fungus (Heath, 1972). Around the haustorial body, many more profiles of endoplasmic reticulum and membrane-bound vesicles are seen in the host cytoplasm than in either the susceptible response or in the intermediate type of resistance (Fig. 24). However, the most striking feature of this type of interaction is the convolution of the extrahaustorial membrane and its association with densely-staining material (Heath and Heath, 1971; Heath, 1972). In the cultivar Queen Anne, where the subsequent rapid necrosis of the invaded cell does not always occur, this material becomes extensive and sometimes shows a periodicity of about 4 nm (Fig. 25), similar to that shown by osmium-fixed phospholipid (Stoeckenius, 1959); the lipid nature of this material is also indicated by the fact that it is etched out of the section during periodate treatment (Fig. 30) (Heath and Heath, 1971). The extensive development of this material on the extrahaustorial membrane coincides with the vacuolation and development of lipid droplets within the haustorium (Fig. 27), both typical signs of fungal senescence. Thus it has been suggested that the abnormality of the extrahaustorial membrane restricts the flow of essential substances to the haustorium and therefore may be a primary determinant in resistance of this cultivar, *when necrosis does not take place* (Heath and Heath, 1971). Significantly, electron microscopy has shown that encasement of these haustoria takes place *after* the

Fig. 17.   The same necrotic host cell (C_1) as seen in Fig. 16 adjacent to an apparently living (at the time of fixation) mesophyll cell (C_2). Both cells contain electron-opaque appositions (A) against their cell walls. Membranous whorls (arrows) are common in the cytoplasm of the living cell. ($\times$ 13,000) (From Skipp *et al.*, 1974. Reproduced by permission of the National Research Council of Canada from *Can. J. Bot.* **52**, 2615-2620.)

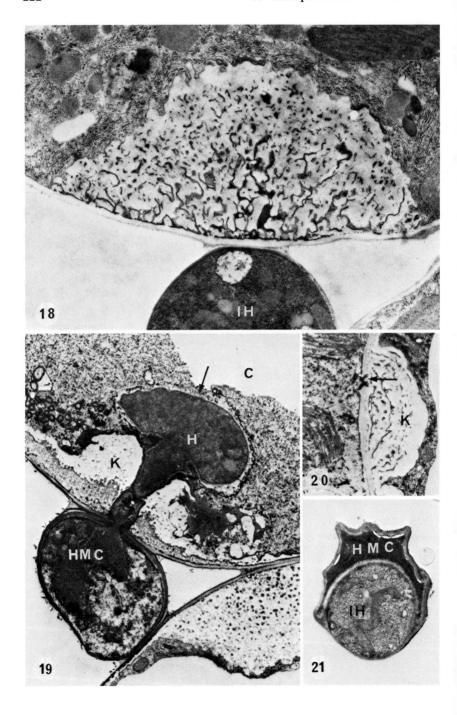

haustorium appears senescent (Fig. 27) and therefore probably has little or no role in determining resistance.

The encasement of the haustoria in cowpea cultivar Queen Anne (Fig. 26) results from the continued growth of the collar of electron-lucent material initially formed around the haustorial neck. Since the investigation of this phenomenon represents the only detailed account of the development of haustorial encasements, it will be described in some detail. As the encasement grows, the adjacent host cytoplasm acquires numerous vesicles which disappear once the haustorium is fully enclosed (Fig. 27). The smaller vesicles are ribosome-studded while the larger have fewer ribosomes, but show more clearly a tripartite structure similar to that of the plasmalemma surrounding the encasement (Fig. 28). The most obvious interpretation from these observations is that precursors of encasement material accumulate in dilated portions of the endoplasmic reticulum whose membranes then undergo transformation to allow fusion with the host plasmalemma. Such fusion results in the release of the content of the vesicles to the encasement. Endoplasmic reticulum involvement in polysaccharide synthesis has been suggested for other systems (Chrispeels, 1976) but normally such membrane transformations occur across the dictyosome (Morré, 1975); however, during formation of the encasement, there is usually no obvious increase in dictyosome number or activity. Periodic acid–silver hexamine (PASH) staining to locate areas rich in polysaccharide does not stain the contents of these vesicles (or Golgi vesicles) although the encasement stains strongly (Figs. 29

Fig. 18.   An infection hypha (IH) of *Uromyces phaseoli* var. *vignae*, 24 hours after inoculation of a highly resistant host cultivar (Queen Anne). An electron-lucent deposit of material demonstrated by light microscopy to contain callose (i.e., fluoresces under ultraviolet light after aniline blue treatment) has formed in the adjacent cell. Note the abundance of membranous material in the deposit and the rough endoplasmic reticulum in the adjacent host cytoplasm. This response in *not* typical of all highly resistant cultivars. (× 13,600) (From Heath and Heath, 1971.)

Figs. 19–21.   Typical end result of haustorium formation in highly resistant interactions.

Fig. 19.   The first haustorium formed in the host-rust combination shown in Fig. 18, 28 hours after inoculation. Host cell (C), haustorium (H), and haustorial mother cell (HMC) have all disorganized and there is no sign of the electron-lucent extrahaustorial matrix where the extrahaustorial membrane has disappeared (arrow). Note the callose-containing collar (K) around the haustorial neck with its more electron-opaque component next to the host cell wall (see also Figs. 23 and 26). A callose-like deposit is also present in the adjacent cell. (× 7,500) (From Heath and Heath, 1971.)

Fig. 20.   Callose-like deposit (K) in an apparently normal cell bordering a necrotic cell similar to that shown in Fig. 19. These deposits commonly form next to plasmodesmata (arrow) between the two cells. (× 9,200) (M. C. Heath, unpublished.)

Fig. 21.   The same host-rust combination as shown in Figs. 18-20. An apparently healthy infection hypha (IH) of a necrotic haustorium and haustorial mother cell (HMC). (× 10,500) (M. C. Heath, unpublished.)

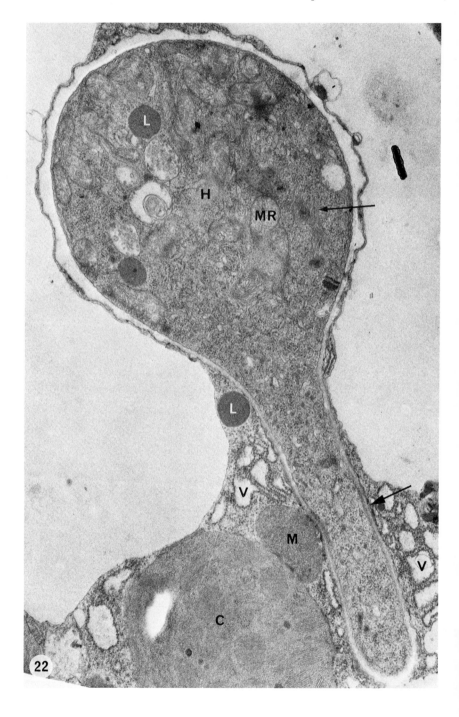

and 30). Thus it seems that if, indeed, these vesicles contribute to the developing encasement, then either the contents change in composition or degree of polymerization after discharge, or the PASH-positive substances of the encasement are secreted in some other way. It is of interest that histochemical evidence suggests that the encasement contains callose (Heath, 1971) which is commonly supposed to be a β-(1 → 3)-glucan; such glucans, however, are resistant to periodate attack except at their terminal residues and thus should be PASH negative. The positive staining of the encasement therefore suggests that either a large number of end groups are present (caused by short-chain lengths or branching of the polysaccharide) or the encasement contains other types of linkages or sugars.

Since the encasement is covered on both sides by the host plasmalemma during development, the final closure over the distal end of the haustorium results in the fusion of the plasmalemma on the outer side of the encasement to form a continuous membrane. Consequently that part of the plasmalemma lining the inner side of the encasement, as well as that constituting the extrahaustorial membrane, is excluded from the host cell (Heath and Heath, 1971).

Of the remaining two studies of highly resistant interactions, that between *Puccinia coronata* f. sp. *avenae* and an oat cultivar (Onoe *et al.*, 1976) resembles those described for *Uromyces phaseoli* var. *vignae* in that no disorganization of host cells is seen until the formation of the first haustorium. However, during the prehaustorial phase of growth, electron-opaque deposits resembling those shown in Fig. 17 develop on the inner surfaces of host cell walls adjacent to the fungus. In addition, electron-opaque material (which may not necessarily be the same as that seen within the cell) increases in the intercellular spaces, particularly at the junctions of neighboring cells. Observations of Golgi body frequencies in more than 100 cell sections suggests that the numbers increase, compared with the compatible interaction, and that a high proportion of these Golgi bodies contain electron-opaque material in their cisternae and associated vesicles. Thus either or

---

Figs. 22–25. Types of interaction, prior to necrosis, between *Uromyces phaseoli* var. *vignae* and highly resistant host cultivars.

Fig. 22. Interaction (1) [see Figs. 23–25 for Interaction (2)]. The first-formed D-haustorium (H) in cultivar Dixie Cream. The nuclei have not yet migrated into the haustorium from the haustorial mother cell and the neckband has not fully developed although a region of slightly increased electron opacity can be seen halfway along the haustorial neck (large arrow). Both host and haustorial cytoplasm look relatively normal apart from an accumulation of ribosome-studded vesicles (V) around the haustorial neck, but in this interaction, the haustoria do not develop further and necrosis of both haustorium and invaded cell rapidly occurs (as in Fig. 19). Virus-like particles (see Chapter 7) can be seen in the haustorial body (small arrow). C, chloroplast; L, lipid droplet; M, microbody; MR, mitochondria. (× 26,700) (From Heath, 1972.)

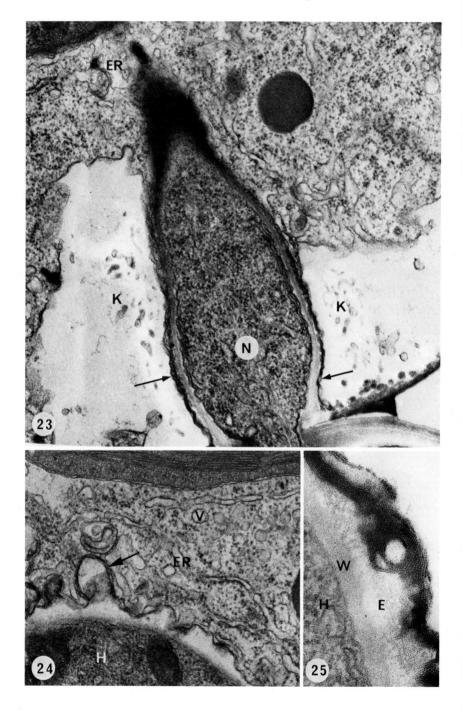

both of the electron-opaque deposits seem to form as a result of Golgi activity. Whether these electron-opaque deposits have a causal role in resistance, however, is unknown but the comparable increase in Golgi bodies seen in a less highly resistant interaction is not accompanied by the presence of this material nor does any deposition of electron-opaque material take place.

The only other ultrastructural study of the more vigorous forms of cultivar resistance is that of El-Gewely *et al.* (1972) using *Melampsora lini*-infected flax. Although completely disorganized host cells were seen near the site of fungal penetration, no early stages in necrosis, nor any haustoria, were observed. The claimed total absence of haustoria, if general for the whole population of infection sites, is most unusual for even highly effective forms of cultivar resistance and thus needs to be substantiated by more extensive investigations.

In all types of cultivar resistance, whether resulting in macroscopic symptoms or not, the characteristic which primarily defines these interactions as being incompatible is a reduction in rate, or complete cessation, of intercellular fungal growth. In those cases where the ultrastucture of intercellular hyphae has been examined, however, the necrosis of adjacent host cells or haustoria seems to have little ultrastructural effect on the adjacent intercellular mycelium (the disorganization of the haustorial cytoplasm seems to be prevented from spreading to the adjacent cell by the haustorial mother cell septum, Fig. 21) (Heath and Heath, 1971; Heath, 1972; Skipp *et al.*, 1974; Van Dyke and Hooker, 1969), even when the fungus develops no further than the stage of the infection hypha. Thus ultrastructural studies have

---

Figs. 23–25.   Interaction (2). [See Fig. 22 for interaction (1).] Haustoria develop normally, but collar formation, and an unusual response by the extrahaustorial membrane, occur prior to necrosis.

Fig. 23.   Oblique section through the neck (N) of the first-formed haustorium in cultivar Calico Crowder. The nuclei have already migrated into the haustorium from the haustorial mother cell and the haustorial cytoplasm looks normal. However, the host has formed a collar (K) of callose-like material around the region of penetration. Note how a fold of the host plasmalemma can be traced between the collar and the haustorial neck (arrows) illustrating that the collar material is not deposited directly onto the fungal wall. The fenestrated sheet of endoplasmic reticulum (ER) covering the upper part of the haustorial neck is also found in susceptible hosts at this stage of haustorium development. (× 41,300) (From Heath, 1972.)

Fig. 24.   A haustorial body (H) from the same type of interaction as shown in Fig. 23. The adjacent host cytoplasm contains abundant rough endoplasmic reticulum (ER) and membrane-bound vesicles (V). Note the convolutions, and increase in electron opacity of the extrahaustorial membrane (arrow). (× 42,000) (From Heath, 1972.)

Fig. 25.   Electron-opaque material associated with the extrahaustorial membrane at a later stage of development than that shown in Fig. 24. Note the myelin-like configurations with a periodicity of about 4 nm. H, haustorial body; W, haustorial wall; E, extrahaustorial matrix. (× 97,600) (From Heath and Heath, 1971.)

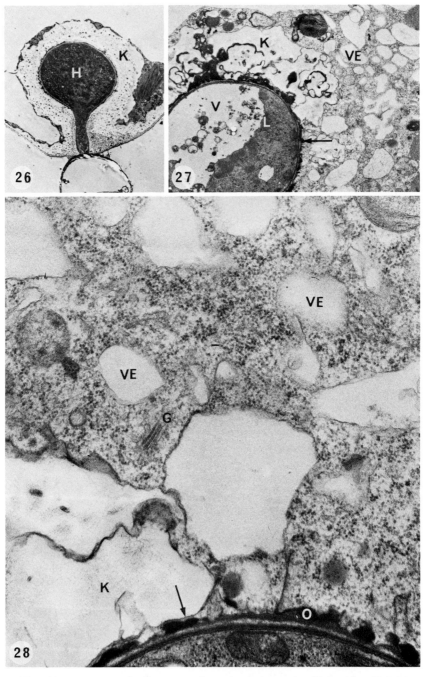

Figs. 26–30. (continued). Encasement formation in interaction (2) (see Figs. 23-25) in instances where necrosis does *not* take place (cultivar Queen Anne).

provided no clear answers to the question of why the fungus stops growing in resistant plants, other than to echo the conclusions from some light microscope studies and transplantation experiments (Allen, 1923a; Chakravarti, 1966; Thatcher, 1943) that hyphal death is not the prime reason.

2. Nonhost Resistance

As mentioned earlier, the resistance of nonhost plants is commonly characterized by the absence of haustoria and the severe restriction of fungal growth. This generalization seems to hold true for urediospore and aeciospore inoculations (Gibson, 1904) and although signs of direct inhibition of infection structure development within the leaf have been reported (Heath, 1974a; Heath, 1977; Leath and Rowell, 1966), in many cases infection hyphae develop normally and produce at least one haustorial mother cell (Gibson, 1904; Heath, 1974a; Heath, 1977); thus the absence of haustoria seemingly cannot be explained by the inhibition of intercellular growth. Since nonhost resistance is the most effective and widespread form of resistance in nature, it is unfortunate that there have been only two published and (to our knowledge) one unpublished ultrastructural investigations of this phenomenon, and all involve either *Uromyces phaseoli* var. *vignae* (cowpea rust, Heath, 1972; Heath, 1974a) or the closely related *U. phaseoli* var. *typica* (bean rust, M. C. Heath, unpublished). In the nonhost responses to *U. phaseoli* var. *vignae,* ultrastructural evidence was found for at least three mechanisms capable of inhibiting haustorium formation (Heath, 1972; Heath, 1974a). In *Phaseolus* spp., the absence of haustoria seems related to the deposition of electron-opaque material within, and upon the French bean cell walls adjacent to the infection hypha; these deposits contain silicon (M. C. Heath, unpublished) and are especially pronounced next to haustorial mother cells (Fig. 31). They are absent, however, in rare cases when haustoria do form (Fig. 35). In *Vicia faba* (broad bean), the lack of haustoria

---

Fig. 26.   A vacuolate (out of plane of section) but otherwise normal haustorium (H) completely encased in callose-containing material (K). (× 3,100) (From Heath and Heath, 1971.)

Fig. 27.   A developing encasement (K). Note that signs of senescence [i.e. the vacuole (V) and lipid droplets (L)] are already present in the haustorial body and the electron-opaque material on the extrahaustorial membrane (arrow) has become extensive. The cytoplasm is full of vesicles (VE) which disappear as soon as the encasement is completed (see Fig. 26). (× 6,500) (M. C. Heath and I. B. Heath, unpublished.)

Fig. 28.   Detail of the edge of a developing encasement (K) showing the way in which the host plasmalemma lines the inner surface of the encasement (arrow). The large vesicles (VE) in the cytoplasm have fewer ribosomes associated with their membranes than the smaller vesicles shown in Fig. 29. Note the scarcity of Golgi bodies (G) and the abundant electron-opaque material on the extrahaustorial membrane (O). (× 30,200) (From Heath and Heath, 1971.)

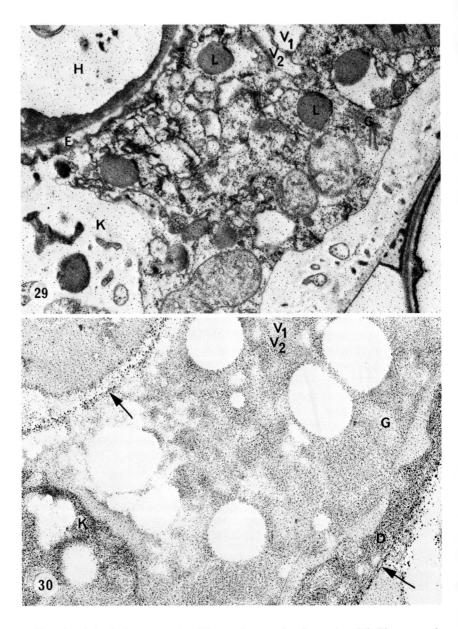

Fig. 29. A developing encasement (K) around a vacuolate haustorium (H). The surrounding cytoplasm is full of irregular ribosome-studded vesicles (V_1, V_2). The uniformly distributed dark granules over the section are caused by treatment with periodic acid–silver hexamine (PASH) but prior irradiation in the electron microscope has eliminated the normal effect of this treatment (i.e., heavy metal staining has been retained and polysaccharides have not been oxidized). This unexplained effect of irradiation allows direct comparison of adjacent serial sections "before and after" PASH treatment. E, extrahaustorial matrix; G, Golgi body; L, lipid droplets. (× 24,100) (From Heath and Heath, 1971.)

appears to be explained, at least in part by the ease with which the mother cell becomes detached from the nonhost cell wall. The electron microscope shows that this seems to be brought about through the separation of the outer fungal wall layer, which is well attached to the nonhost cell, from the remaining layers of the fungal wall (Fig. 32). That this might have resulted from some activity on the part of the plant is suggested by the presence of the cell nucleus near the potential site of penetration and by an unusual accumulation of vesicles and tubules, some obviously continuous with the plasmalemma, in this region (Fig. 33). Such responses are not seen in other nonhost plants nor in susceptible interactions. In contrast, the absence of haustoria in *Pisum sativum* (pea) correlates with the complete disorganization of the infection hypha just before or just after haustorial mother cell formation; no ultrastructural changes can be detected in the cytoplasm of the surrounding plant cells (Fig. 34).

A particularly interesting observation, when considering the basis of host–parasite specificity (see below), is that in the rare cases where haustoria of *Uromyces phaseoli* var. *vignae* do form in *Phaseolus vulgaris* and *Pisum sativum*, haustorium development appears identical to that in the susceptible host (Fig. 35) and the invaded cell initially shows no obvious adverse reaction other than a possible increase in polyribosomes in *P. sativum* (Heath, 1972, 1974a). However, rapid disorganization of haustorium and invaded cell cytoplasms soon takes place but, as for cultivar resistance, the infection hypha remains ultrastructurally normal.

As an exception to the general rule of haustorium absence in nonhost plants, inoculation of the nonhost *Vigna sinensis* (cowpea) with *Uromyces phaseoli* var. *typica* commonly results in the fungus producing one, or, at the most, two haustoria (M. C. Heath, unpublished). These haustoria begin to develop normally, but extremely rapid and complete disorganization of the invaded cell cytoplasm occurs before the migration of the fungal nuclei from the haustorial mother cell (Figs. 36 and 37). However, although the haustoria do not complete their development, they remain ultrastructurally normal for some time. This situation is unique, so far as we know, among the types of host and nonhost responses which involve rapid necrosis of the invaded cell.

Fig. 30. Nonirradiated section adjacent to that shown in Fig. 29. The host and haustorial walls (arrows), the encasement (K), and the callose-like deposit (D) on the host wall have all reacted strongly. Some of the vesicles (V_1, V_2) and the Golgi body (G) seen in Fig. 29 can be located but show no increase in staining relative to the surrounding cytoplasm. PASH treatment has caused the lipid droplets and extrahaustorial matrix (or, more likely, the membrane with its electron-opaque deposits) to be etched away leaving holes which have allowed the section to expand during examination. ($\times$ 19,500) (From Heath and Heath, 1971.)

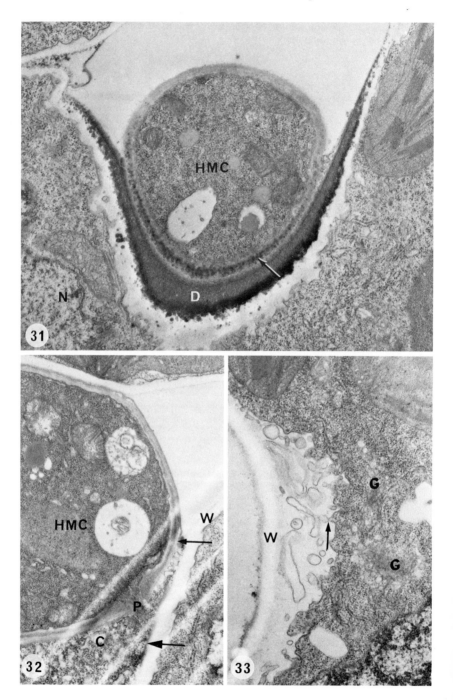

C. Overview

Obviously many more ultrastructural studies of different types of host and nonhost resistance to rust infection are badly needed. While not solving the fundamental problems of *why* such resistance occurs, the studies made to date have nonetheless contributed some highly significant information. First, there is the diversity of ultrastructural changes observed among different host–parasite combinations. Even where some particular feature is found in more than one combination (e.g., the development of a granular or fibrillar extrahaustorial matrix in some types of cultivar resistance), it is rarely found in all and any similarity in this one feature is commonly offset by a lack of similarity in others. In particular, these studies seem to confirm the conclusion reached from nonultrastructural work by a few researchers that there are many ways in which a cell may die (Wood, 1973; Littlefield, 1973; Heath, 1976a); not only can the speed of disorganization differ between responses, but also the way in which this disorganization occurs and the immediate effect on the haustorium may vary. Thus the ultrastructural evidence does not so far support Stakman's (1914) hypothesis that the different types of resistant responses to rust differ only in a matter of degree, and it seems highly unlikely that there will be one general mechanism for resistance which can be applied to all types of interaction.

Not only is there diversity among the ultrastructural features of resistance to different rusts, but diverse responses may be induced by the same rust in different plants. The studies of the resistance of four host cultivars and three nonhost species to a single race of *Uromyces phaseoli* var. *vignae* (Heath and Heath, 1971; Heath, 1972, 1974a) suggest that there are many stages during

---

Figs. 31–33. Nonhost responses to *Uromyces phaseoli* var. *vignae*.

Fig. 31. An obliquely sectioned deposit (D) of material in *Phaseolus vulgaris* next to the first-formed haustorial mother cell (HMC). Note the electron-opaque material (arrow) in that part of the fungal wall closest to the plant cell and the presence of the plant nucleus (N) and abundant endoplasmic reticulum in this region. In most infection sites, no haustorium is formed in this rust–plant interaction. (× 21,900) (From Heath, 1974a.)

Fig. 32. The first-formed haustorial mother cell (HMC) in *Vicia faba* which has lost contact with the plant cell wall (W). Note the broken end of the outer fungal wall layer (small arrow) and what appears to be a portion of the fungal wall (large arrow) still attached to the plant wall. The penetration peg (P) has burst and cytoplasmic material (C) can be seen in the intracellular space. As in *P. vulgaris* (Fig. 31), haustoria rarely form in this nonhost plant. (× 21,900) (From Heath, 1974a.)

Fig. 33. A *Vicia faba* cell adjacent to a detached haustorial mother cell (not shown) similar to that shown in Fig. 32. Note the accumulation of tubules and vesicles in this region, some of which (arrow) seem to be continuous with the plant plasmalemma. A number of Golgi bodies (G) are also present in this region. The electron-opaque material on the outer surface of the plant wall (W) may represent the region where the haustorial mother cell had been attached. (× 21,900) (From Heath, 1974a.)

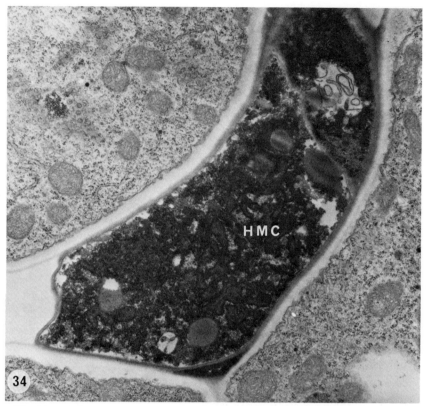

Fig. 34.   Nonhost response to *Uromyces phaseoli* var. *vignae*. The first-formed haustorial mother cell (HMC) in *Pisum sativum*. The fungal cytoplasm has completely disorganized but the ultrastructure of the surrounding host cells is indistinguishable from that in uninfected leaves. (× 17,600) (M. C. Heath, unpublished.)

the infection process whereby an interaction between plant and parasite may result in the expression of resistance. It has been suggested that these stages represent "switching points" where the outcome of the interaction determines the subsequent progress of infection (Fig. 1, see page 209) (Heath, 1974a). Thus susceptibility depends on the "correct" response at *every* "switching point" while resistance theoretically requires an "incorrect" response at only one of these stages. It remains to be seen whether those examples of cultivar

Fig. 35.   Nonhost response to *Uromyces phaseoli* var. *vignae*. One of the few instances where a haustorium has formed in *Phaseolus vulgaris*. There is little or no wall deposit (see Fig. 31) adjacent to the haustorial mother cell and the ultrastructure of both plant cell and haustorium appear similar to those in compatible interactions. However necrosis of both plant cell and haustorium occurs soon after haustorium formation. (× 25,200) (From Heath, 1972.)

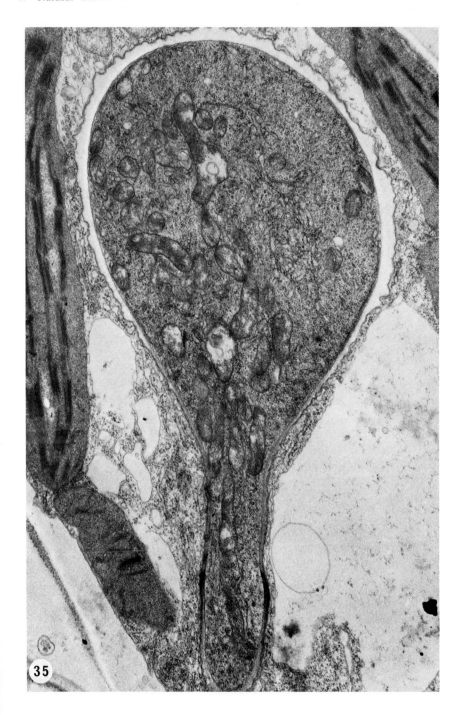

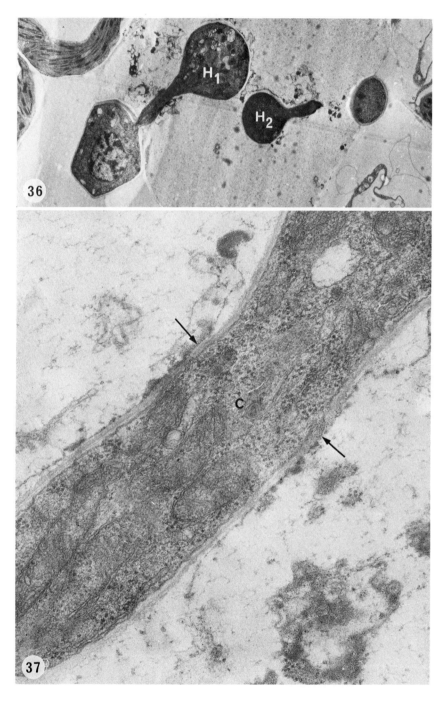

Figs. 36–37. Nonhost response of *Vignae sinensis* to *Uromyces phaseoli* var. *typica.*

resistance to *U. phaseoli* var. *vignae* which are manifested at different "switching points" represent the involvement of different major genes for resistance. One would expect this to be so since Littlefield (1973) has shown for *Melampsora lini* that different genes for resistance in the host result in recognizably different types of interactions when examined with the light microscope. A similar situation has also been shown to exist for resistance to powdery mildew (Ellingboe, 1972; Lupton, 1956).

It is important for our understanding of the mechanisms of resistance to know whether necrosis of the haustorium-containing cell occurs before or after the death of the haustorium (Thatcher, 1943). It has recently been suggested from experiments involving the treatment of susceptible infections with blasticidin S or nickel nitrate that host necrosis is induced by the leakage of fungal products after the fungus has been killed (Kiraly *et al.*, 1972; Barna *et al.*, 1974). However, Kim *et al.* (1977) have disputed the validity of this interpretation of the results, and have carried out similar experiments where histological observation did not support this conclusion. It is possible that the situation may differ in different types of resistance, but it is perhaps significant that in *none* of the ultrastructural studies carried out so far has the invaded cell been seen to become necrotic *after* the death of the haustorium. In all cases either the death of the fungus and host cell appear simultaneous, or the host cell disorganizes first. This strongly supports the conclusion by Kim *et al.* (1977) that, whatever its role in the limitation of fungal growth, host cell death is a consequence of some activity of the living fungus and not merely a response to the toxic products of fungal death.

Another general conclusion that can be made from these ultrastructural studies of resistance is the lack of importance of haustorium encasement as a resistance mechanism (Coffey, 1976; Heath and Heath, 1971), even though encased haustoria have been seen with the light microscope in a number of incompatible interactions (e.g., Heath, 1971; Littlefield and Aronson, 1969; Thatcher, 1943) and in some (Heath, 1971; Thatcher, 1943) they seem to be the most conspicuous feature of the interaction. In the two examples examined with the electron microscope, encasement occurs *after* the haustorium shows signs of starvation (Heath and Heath, 1971) or changes are seen in the

---

Fig. 36. The first (H_1) and second (H_2) (verified by serial sections of the infection hypha) haustoria formed by a single, bifurcated, infection hypha in a single cell. Note the complete disorganization of the plant cell contents. ($\times$ 5,200) (M. C. Heath, unpublished.)

Fig. 37. Detail of the neck of H_1 shown in Fig. 36. The haustorial cytoplasm (C) looks normal and fragments of the extrahaustorial membrane can be seen attached to what seems to be the neckband (arrows). Note the uniform width of the fungal wall on both sides of the neckband and the absence of any signs of the extrahaustorial matrix where the extrahaustorial membrane is lacking (see Chapter 4, Section II,A,1,c, for the significance of this observation). ($\times$ 57,000) (M. C. Heath, unpublished.)

appearance of the extrahaustorial matrix (Coffey, 1976). Thus encasement in the resistant plant seems to represent the same nonspecific host response to impaired functioning of the haustorium (Coffey, 1976) as it does in compatible interactions (see Chapter 4, Section II,C,2).

Finally, it is vital to our understanding of the nature of host specificity in the rusts, to know whether the invaded cell inherently recognizes the haustorium as non-self (and therefore is likely to react against it unless the fungus in some way negates this response), or whether there is no specific non-self recognition phenomenon in vascular plants. The fact that apparently normal haustoria of *Uromyces phaseoli* var. *vignae* can develop in nonhost plants without any initial response of the invaded cell other than that seen in the compatible association (Heath, 1972, 1974a) strongly suggests that haustorium formation does not require any special adaptation on the part of the fungus towards its host, and that host (i.e., species) specificity is not determined at the level of haustorium development. This supports the evidence from protoplast fusion in higher plants (see Bushnell, 1976) and cytoplasmic exchange between two taxonomically distinct fungi (Hoch, 1977) that nonself recognition may be poorly, if at all, developed in organisms with cell walls. If this hypothesis is correct, the fact that highly resistant cultivars of the host respond adversely to haustorium formation whereas nonhosts do not (Heath and Heath, 1971; Heath, 1972) suggests an adaptation on the part of the plant towards intolerance of processes it normally would not react to. Correspondingly, the adaptation of the fungus to its susceptible host must therefore not be in the realm of haustorium formation, but rather in whatever processes are involved in the sudden subsequent cell necrosis which eventually occurs in nonhost plants. Unfortunately this hypothesis is based on information for only one rust species and light microscope studies of nonhost responses to other rusts suggest that it may not be universally applicable (Heath, 1977). It is hoped that future ultrastructural studies will provide more information on this interesting aspect of rust resistance.

II. RESISTANCE ELICITED BY
SYSTEMIC FUNGICIDES

Ultrastructural studies on the effects of systemic fungicides on rust infection are of theoretical interest, not only to supply information on the mode of action of the fungicide, but to also determine similarities between this "artificial" resistance and that conditioned "naturally" by the plant genome. Unfortunately there have been few such studies, and the fungicide chosen in all but one case has been oxycarboxin, 5,6-dihydro-2-methyl-1,4-oxathiin-3-carboxanilide 4,4-dioxide, previously designated 2,3-dihydro-6-methyl-5-

phenyl-carbamoyl-1,4-oxathiin 4,4-dioxide. This compound (also known as DCMOD®, Uniroyal Plantvax®, Uniroyal F461®) inhibits mitochondrial respiration close to the site of succinate oxidation and is translocated apoplastically through the plants (see references in Pring and Richmond, 1976).

Although Bisiach and Locci (1972) found by light and scanning electron microscopy that oxycarboxin treatment of *Phaseolus vulgaris* prior to inoculation resulted in deleterious effects on urediospore germination, germ tube growth, and appressorium formation of *Uromyces phaseoli* var. *typica,* Pring and Richmond (1976), using light microscopy of the same rust, only observed a reduction in germ tube length and found that infection hyphae, but no haustoria, developed within the leaf. Transmission electron microscope studies on the effect of oxycarboxin on *U. phaseoli* var. *typica* have been limited to infections treated with the fungicide at the flecking stage of development (Pring and Richmond, 1976), and these suggest that in this situation the haustoria are the first fungal structures to show signs of damage. This is rather surprising since the intercellular hyphae should be the first fungal structure reached by a fungicide translocated in the apoplast. Pring and Richmond explain this effect on the haustoria by suggesting that they may be sufficiently strong metabolic sinks to enable them to accumulate the fungicide from the small amounts entering the symplast; alternatively the increased permeability of infected cells (see Chapter 4, Section II,C,1) may enable larger quantities of oxycarboxin than expected to enter the invaded cell and thus reach the haustorium. These authors also suggest that entry into the haustorium may take place through the haustorial neck, but since this conclusion is partially based on the questionable absence of fungal wall at the point of host wall penetration (see Chapter 4, Section II,A), this hypothesis must be viewed with caution.

As one might expect from the mode of action of this fungicide, swelling and disruption of the haustorial mitochondria are conspicuous features of the fungal response (Fig. 38) and can be seen 20–24 hours after fungicide application to the soil (Richmond, 1975; Pring and Richmond, 1976). Later vacuolation and lipid accumulation occur as they do in naturally resistant interactions where the haustorium is presumed to be starved (see Section I of this chapter). More unexpectedly, for a response to a compound which is supposed to act on the fungus alone, the extrahaustorial membrane also fragments (Fig. 38), although Pring and Richmond (1976) suggest that this may represent a "weakening" of the membranes in living cells so that they fragment on fixation. By 6 days after treatment the haustoria are shrunken and electron-opaque and they eventually become encased (Fig. 39), again resembling the response to haustoria in some naturally resistant host cultivars

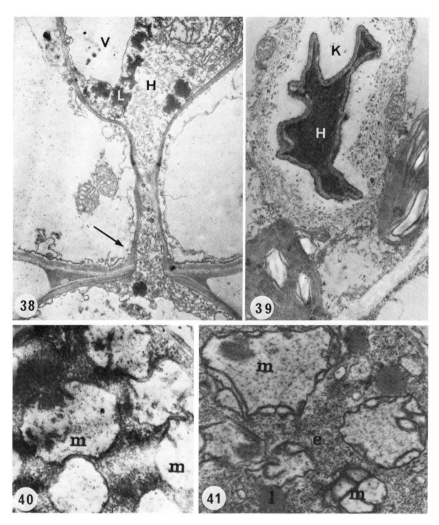

Figs. 38–41. "Resistance" induced in compatible interactions by systemic fungicides.

Fig. 38. A haustorium (H) of *Uromyces phaseoli* var. *typica*, 9 days after treatment of a susceptible host plant with oxycarboxin. The haustorial body contains a vacuole (V) and lipid droplets (L), both signs of senescence. Note the fragmentation of the host plasmalemma (e.g. arrow). (× 12,800) (From Pring and Richmond, 1976.)

Fig. 39. A disorganized haustorium (H) of *U. phaseoli* var. *typica* completely encased in callose-like material (K) (compare with Fig. 26 of this chapter and Figs. 80–81 in Chapter 4) 6 days after treatment of the infected susceptible plant with oxycarboxin. (× 9,500) (From Pring and Richmond, 1976.)

Fig. 40. Mitochondria (m) of a haustorium of *Puccinia coronata* f. sp. *avenae*, 24 hours after application of oxycarboxin to the soil around the susceptible host plant. (× 23,400) (From Simons, 1975.)

(see Section I of this chapter) and in some compatible interactions (see Chapter 4, Section II,C,2), and probably representing a similarly non-specific, "walling-off " response to the necrotic haustorium. Encasement was also found to be the most conspicuous feature of oxycarboxin-treated infections of *Puccinia horiana* (Kajiwara and Takahashi, 1975). Following haustorial necrosis in *Uromyces phaseoli* var. *typica* infections, the intercellular hyphae undergo a morphologically similar disintegration (Pring and Richmond, 1976).

In a comparative study of two different systemic fungicides, oxycarboxin has also been seen to induce enlargement and disorganization of haustorial mitochondria in *Puccinia coronata* f. sp. *avenae,* this time within 12–24 hours after soil treatment (Fig. 40) (Simons, 1975); effects on the extra-haustorial membrane, however, were not noted. By comparison, 12 hours after benomyl [methyl 1-(butylcarbamoyl)-2-benzimidazolecarbamate] treatment there are no obvious changes in haustorial ultrastructure, but by 24 hours nuclear membranes and plasmalemmas appear damaged or missing, and the enlarged mitochondria often show curious loop configurations of the cristae (Fig. 41) (Simons, 1975), a phenomenon also seen in benomyl-treated *Botrytis fabae* (Richmond and Pring, 1971). Presumably the morphological effects of these two fungicides differ because of their different modes of action; however, the effectiveness of benomyl as a fungicide is mainly due to the plant-derived decomposition product, methyl 2-benzimidazolecarbamate (MBC) for which the only clearly understood mode of action is specific binding to fungal tubulin, the structural protein of microtubules (Davidse and Flach, 1977). How this relates to the ultrastructure changes seen in haustoria remains to be determined.

Fig. 41.   Mitochondria (m) of a haustorium of *P. coronata* f. sp. *avenae,* 24 hours after application of benomyl to the soil around the susceptible host plant. Note the curious looplike configurations of the mitochondrial cristae. e, endoplasmic reticulum; l, lipid droplet. ($\times$ 24,800) (From Simons, 1975.)

7

Parasites of Rust Fungi

I. FUNGI

Many species of rust fungi, in various stages of their life cycle, have been reported to be hosts for the hyperparasitic fungus *Darluca filum*. Commonly this fungus is observed as pycnidia within uredia of the rust (Fig. 1). In the uredial stage of *Puccinia graminis*, the infection is restricted to urediospores, with no infection observed in the basal cells of the sorus (Carling *et al.*, 1976). Penetration of the urediospores by *Darluca filum* occurs at random sites over the spore, not through germ pores, and appears to be dependent largely on the enzyme-producing capacity of the hyperparasite (Figs. 2 and 3). The penetrating hypha is not constricted at the point of entry; it proceeds to form a septate, intracellular hypha in the urediospore. Once penetrated, the host cell cytoplasm undergoes rapid disorganization. Numerous other genera of Deuteromycetes have been reported as parasitic on Uredinales (see von Shroeder and Hassebrauk, 1957, for review), but no ultrastructural investigations have been made of such associations.

Fig. 1.   Pycnidia (PY) of *D. filum* occur within uredia of *P. graminis*. U, urediospore. (× 750)

Fig. 2.   Growth of *D. filum* hyphae (H) among urediospores (U) in a uredium. Points of intimate contact (arrows) are sites of potential hyphal penetration. (× 1500)

Fig. 3.   Urediospore of *P. graminis* showing penetration by septate hypha (H) of *D. filum*. Note the lack of marked constriction of the parasitic hypha at the point of urediospore penetration, also the clear continuity of cell wall (CW) between the inter- and intracellular portions of the hyperparasite. The cytoplasm (CY) of the host is disorganized and lacks normal organelles. (× 12,750)

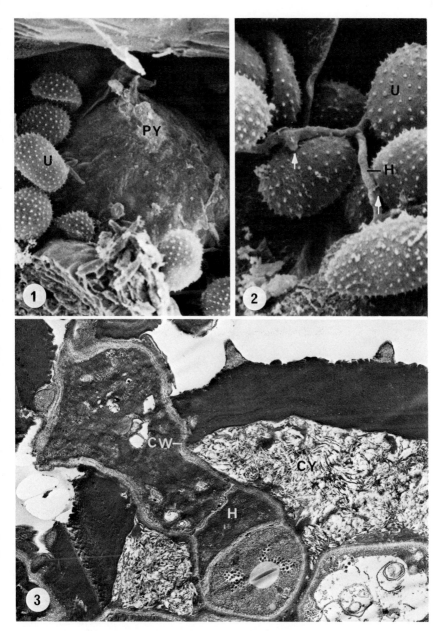

Figs. 1–3.   Infection of *Puccinia graminis* by the hyperparasite, *Darluca filum*. (From Carling *et al.*, 1976.)

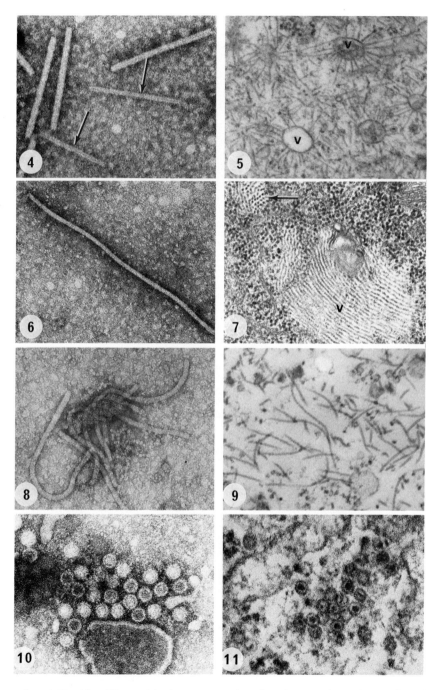

Figs. 4–11. Virus-like particles in *Uromyces phaseoli* var. *vignae*. (From McDonald and Heath, 1978. Reproduced by permission of the National Research Council of Canada from *Can. J. Bot.* **56,** 963–975.)

II. VIRUSES

Although it is now recognized that viruses occur widely in fungi (Lemke, 1976), there have been relatively few reports of virus-like particles (VLPs) in ultrastructural studies of the rusts. In one of the first of such reports, Yarwood and Hecht-Poinar (1973) claim to have found rod-shaped particles in palladium-shadowed extracts of tissues infected with various rusts and powdery mildews. Extracts from *Uromyces phaseoli* var. *typica* infections reacted positively with Tobacco Mosaic Virus (TMV) antiserum and thus the authors suggest that TMV-like viruses occur in several members of these two distinct groups of pathogens. However the single published micrograph shows particles predominantly smaller than TMV and inconsistencies between the observation of VLPs and positive reactions with TMV antiserum suggest that the apparent serological similarity may have been due to low levels of TMV contamination. It is also unclear from this work whether the VLPs are really inside the fungus or whether they are surface contaminants acquired by passage through an infected host.

In a more detailed investigation of one of the rusts (*Uromyces phaseoli* var. *vignae*) used by Yarwood and Hecht-Poinar (1973), McDonald and Heath (1978) show that three rod-shaped VLPs (a short, rigid rod 260 × 12 nm, Fig. 4; a long, flexuous rod 660 × 10 nm, Fig. 6; and a long, wider flexuous rod 740 × 16 nm, Fig. 8) and a 35 nm diameter spherical VLP (Fig. 10) can be found in negatively stained extracts of infected tissue and extracts of isolated, washed urediospores and germ tubes. The size, distribution, and stability of the rods are affected by choice of stain (compare Figs. 8 and 12), pH, and presence of salt, but none resemble TMV morphologically;

Fig. 4.   Two short, rigid rods (arrows) adjacent to some Tobacco Mosaic Virus particles. Negatively stained with 1% ammonium molybdate, pH 7.0. (× 96,500)

Fig. 5.   Thin section containing what seem to be the rigid rods shown in Fig. 4. Characteristically these radiate from membrane-bound vesicles (V) which contain groups of smaller vesicles. Similar vesicles can also be found in the fungal cytoplasm unassociated with virus-like particles. (× 57,200)

Fig. 6.   Narrow, flexuous rod negatively stained with 1% ammonium molybdate, pH 7.0. (× 96,500)

Fig. 7.   Longitudinal (V) and transverse (arrow) sections of what seem to be aggregates of the narrow, flexuous rods shown in Fig. 6. (× 44,800)

Fig. 8.   Tangled cluster of wide, flexuous rods. Negatively stained with 1% ammonium molybdate, pH 7.0. (× 81,300)

Fig. 9.   Thin section containing what may be the wide, flexuous rods shown in Fig. 8. (× 57,200)

Fig. 10.   Spherical particles negatively stained with 2% phosphotungstic acid, pH 7.0. (× 96,500)

Fig. 11.   Spherical virus-like particles in thin section. (× 97,300)

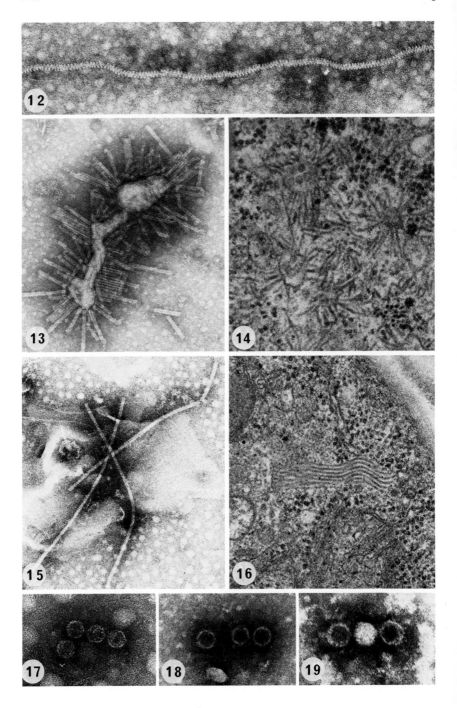

McDonald and Heath suggest that the short rod is equivalent to that seen by Yarwood and Hecht-Poinar (1973).

Transmission electron microscopy shows that morphologically comparable VLPs can be recognized at various stages of fungal growth from urediospore germination to pustule formation (Figs. 5, 7, 9, 11), and in advanced pycnial infections, but the abundance of each type, particularly the rods, seems to differ at different stages of fungal development. This observation, together with the fact that a West African isolate of this same rust contains the same four types of VLP's found in the North American isolate, suggests a well-tuned and long-standing fungus-virus (assuming the VLPs to be viruses) relationship in this rust.

A preliminary survey of negatively stained extracts, and thin sections, of uredial and telial stages of other rusts (McDonald and Heath, unpublished) has shown all four types of cowpea rust VLPs in single isolates of *Uromyces phaseoli* var. *typica* and *Puccinia helianthi* race 2 (Figs. 15, 16, and 18) and the short rods and spherical VLPs in *P. sorghi* (Figs. 13, 14, and 17). Spherical particles only were found in *P. graminis* var. *tritici* race 56 (Fig. 19) and field collections of *Coleosporium solidaginis* and *P. malvacearum*. These spherical VLPs may eventually be found to be ubiquitous in the Uredinales since they can be discerned in published micrographs of thin sections of the uredial stages of *Hemileia vastatrix* (Rijo and Sargent, 1974) and different isolates of *P. phaseoli* var. *typica* (Hardwick *et al.*, 1971) and *P. helianthi* (Coffey *et al.*, 1972a), and have also been obtained from leaf extracts of plants infected with *P. malvacearum* (telial stage), *P. striiformis* (uredial stage) and *P. suaveolens* (pycnial and uredial stages) (Lecog *et al.*, 1974). A spherical VLP has also been seen in negatively stained extracts and thin sections of several variants of *P. graminis* var. *tritici* grown in axenic culture (Mussel *et al.*, 1973; Rawlinson and Maclean, 1973). Although the latter VLP was found not to react in agar gel-diffusion tests against antiserum preparations from other fungal viruses (Rawlinson and Maclean,

---

Fig. 12. Wide, flexuous rod in *Uromyces phaseoli* var. *vignae* with a uniformly unwound helix (compare with Fig. 8). Negatively stained with 2% phosphotungstic acid. (× 96,500) (From McDonald and Heath, 1978. Reproduced by permission of the National Research Council of Canada from *Can. J. Bot.* **56**, 963–975.)

Figs. 13 and 14. Negatively stained whole-mount and thin section, respectively, of short, rigid rods in *Puccinia sorghi*. Note their similarity to those shown in Figs. 4 and 5. (Fig. 13, × 77,800; Fig. 14, × 59,200) (J. G. McDonald and M. C. Heath, unpublished.)

Figs. 15 and 16. Negatively stained whole-mount and thin section, respectively, of narrow, flexuous rods in *P. helianthi*. Note their similarity to those shown in Figs. 6 and 7. (Fig. 15, × 86,500; Fig. 16, × 63,700) (J. G. McDonald and M. C. Heath, unpublished.)

Figs. 17, 18, and 19. Negatively stained, spherical particles from *P. sorghi*, *P. helianthi*, and *P. graminis* f. sp. *tritici* respectively. Again, note their similarity to those shown in Fig. 10. (Figs. 17 and 18, × 136,700; Fig. 19, × 186,900) (Courtesy of J. G. McDonald, unpublished.)

1973), all these spherical VLPs observed in the rusts seem to fit morphologically into the rather characteristic group of double-stranded RNA viruses which have been found widely in the basidiomycetes, ascomycetes and deuteromycetes (Lemke, 1976). In contrast, none of the few examples of rod-shaped VLPs found in other fungi (Dielman-Van Zaayen et al., 1970; Huttinga et al., 1975; Ushiyama and Hushioka, 1973) morphologically resemble those reported for U. phaseoli var. vignae, and the seemingly erratic distribution of these latter VLPs among other rusts, together with the difference in abundance of each type during different stages of development would seem to make these VLPs worthy of further study.

Finally, it should be pointed out that none of these rust VLPs have been shown unequivocally to be viruses. When, or if, this is proven, their effect on the metabolism of the fungus and on the rust's relationship with its host plant will become an area which will need to be clarified.

Bibliography

Abu-Zinada, A. A. H., Cobb, A., and Boulter, D. (1975). An electron-microscopic study of the effects of parasite interaction between *Vicia faba* L. and *Uromyces fabae*. *Physiol. Plant Pathol.* **5**, 113–118.

Adams, J. F. (1919). I. Rusts on conifers in Pennsylvania. II. Sexual fusions and development of the sexual organs in the Peridermiums. *Bull. Penn. Agric. Exp. Sta.* **160**, 31–77.

Ainsworth, G. C. (1971). "Ainsworth and Bisby's Dictionary of the Fungi." 6th Ed. Commonwealth Mycological Institute, Kew.

Aist, J. R. (1976). Cytology of penetration and infection—fungi. *In* "Physiological Plant Pathology" (R. Heitefuss and P. H. Williams, eds.), pp. 197–221. Springer-Verlag, Berlin.

Akai, S., Fukutomi, M., Kunoh, H., and Shiraiski, M. (1976). Fine structure of the spore wall and germ tube change during germination. *In* "The Fungal Spore, Form and Function" (D. J. Weber and W. M. Hess, eds.), 335–411. John Wiley and Sons, New York.

Alexopoulos, C. J. (1962). "Introductory Mycology." John Wiley and Sons, New York.

Allen, F. H. E., Coffey, M. D., and Heath, M. C. (1979). Plasmolysis of rusted flax: A fine-structural study of the host-pathogen interface (in preparation).

Allen, P. J. (1957). Properties of a volatile fraction from uredospores of *Puccinia graminis* var. *tritici* affecting their germination and development. I. Biological activity. *Annu. Rev. Plant Physiol.* **32**, 385–389.

Allen, R. F. (1923a). A cytological study of infection of Baart and Kanred wheats by *Puccinia graminis tritici*. *J. Agric. Res.* **23**, 131–152.

Allen, R. F. (1923b). Cytological studies of infection of Baart, Kanred and Mindum wheats by *Puccinia graminis tritici* forms III and XIX. *J. Agric. Res.* **26**, 571–604.

Allen, R. F. (1928). A cytological study of *Puccinia glumarum* on *Bromus marginatus* and *Triticum vulgare*. *J. Agric. Res.* **36**, 487–513.

Allen, R. F. (1932a). A cytological study of heterothallism in *Puccinia triticina*. *J. Agric. Res.* **44**, 733–754.

Allen, R. F. (1932b). A cytological study of heterothallism in *Puccinia coronata*. *J. Agric. Res.* **45**, 513–541.

Allen, R. F. (1933a). Further cytological studies of heterothallism in *Puccinia graminis*. *J. Agric. Res.* **47**, 1–16.

Allen, R. F. (1933b). The spermatia of corn rust, *Puccinia sorghi*. *Phytopathology* **23**, 923–925.

Allen, R. F. (1934a). A cytological study of heterothallism in flax rust. *J. Agric. Res.* **49**, 765–791.

Allen, R. F. (1934b). A cytological study of heterothallism in *Puccinia sorghi. J. Agric. Res.* **49**, 1047–1068.

Allen, R. F. (1935). A cytological study of *Puccinia malvacearum* from the sporidium to the teliospore. *J. Agric. Res.* **51**, 801–818.

Amerson, H. V., and Van Dyke, C. G. (1978). The ontogeny of echinulation in uredospores of *Puccinia sparganioides. Exp. Mycol.* **2**, 41–50.

Andrus, C. F. (1931). The mechanism of sex in *Uromyces appendiculatus* and *U. vignae. J. Agric. Res.* **42**, 559–587.

Andrus, C. F. (1933). Sex and accessory cell fusions in the Uredineae. *J. Wash. Acad. Sci.* **23**, 544–557.

Antonopoulos, A. A., and Chapman, R. L. (1976). Morphology of *Cronartium fusiforme* aeciospores: A light and scanning electron microscopic study. *Bot. Gaz.* **137**, 285–289.

Arthur, J. C. (1925). Terminology of Uredinales. *Bot. Gaz.* **80**, 219–223.

Arthur, J. C. (1929). "The Plant Rusts (Uredinales)." John Wiley and Sons, New York.

Arthur, J. C. (1934). "Manual of the Rusts in United States and Canada." Purdue Research Foundation, Lafayette, Indiana.

Barna, B., Ersek, T., and Mashaal, S. F. (1974). Hypersensitive reaction of rust-infected wheat in compatible host–parasite relationships. *Acta Phytopathol. Acad. Sci. Hung.* **9**, 293–300.

Bartnicki-Garcia, S. (1973). Fundamental aspects of hyphal morphogenesis. *Symp. Soc. Gen. Microbiol.* **23**, 245–267.

Beckett, A., Heath, I. B., and McLaughlin, D. J. (1974). "An Atlas of Fungal Ultrastructure." Longman, London.

Bega, R. V. (1960). The effect of environment on germination of sporidia in *Cronartium ribicola. Phytopathology* **50**, 61–69.

Bega, R. V., and Scott, H. A. (1966). Ultrastructure of the sterigma and sporidium of *Cronartium ribicola. Can. J. Bot.* **44**, 1726–1727.

Benedetti, E. L., and Bertolini, B. (1963). The use of phosphotungstic acid (PTA) as a stain for the plasma membrane. *J. R. Microsc. Soc.* **81**, 219–222.

Bisiach, M., and Locci, R. (1972). The *Phaseolus vulgaris–Uromyces appendiculatus* complex. II. Irregular uredospore germination on plants from seeds treated with DCMOD. *Riv. Patol. Veg.* **8**, 127–136.

Blackman, V. H. (1904). On the fertilization, alternation of generations, and general cytology of the Uredineae. *Ann. Bot. (London)* **18**, 323–373.

Bole, B., and Parsons, E. (1973). Scanning electron microscopy of the internal cellular structure of plants. *J. Microsc. (Oxford)* **98**, 91–97.

Bose, A., and Shaw, M. (1971). Sporulation and pathogenicity of an Australian isolate of wheat rust grown *in vitro. Can. J. Bot.* **49**, 1961–1964.

Bossányi, G., and Oláh, G. M. (1972). Mise en évidence des substances de nature polysaccharidique dans la cellule de blé parasitée par *Puccinia graminis* var. *tritici. C. R. Hebd. Seances Acad. Sci., Ser. D.* **274**, 3387–3390.

Bossányi, G., and Oháh, G. M. (1974). Activité phosphatique acide en relation avec la pénétration intracellulaire du parasite *Puccinia graminis* var. *tritici* dans la cellule du blé. *Mycopath. Mycol. Appl.* **54**, 161–171.

Boyce, J. S. (1961). "Forest Pathology." McGraw-Hill, New York.

Boyer, M. G., and Isaac, P. K. (1964). Some observations on white pine blister rust as compared by light and electron microscopy. *Can. J. Bot.* **42**, 1305–1309.

Bracker, C. E. (1967). Ultrastructure of fungi. *Annu. Rev. Phytopathol.* **5**, 343–374.

Bracker, C. E. (1974). The endomembrane system of fungi. *Proc. 6th Int. Congr. Electron Microsc., Canberra* **3**, 558–559.

Bracker, C. E., and Littlefield, L. J. (1973). Structural concepts of host-pathogen interfaces. *In* "Fungal Pathogenicity and the Plant's Response" (R. J. W. Byrde and C. V. Cutting, eds.), pp. 159-318. Academic Press, London.

Bracker, C. E., Ruiz-Herrera, J., and Bartnicki-Garcia, S. (1976). Structure and transformation of chitin synthetase particles (chitosomes) during microfibril synthesis *in vitro*. *Proc. Natl. Acad. Sci. U.S.A.* **73**, 4570-4574.

Brandenburger von, W. (1971). Rostpilze aus Ligurien/Italien. *Nova Hedwigia Z. Kryptogamenkd.* **21**, 137-191.

Brandenburger von, W., and Schwinn, F. J. (1971). Uber Oberflächenfeinstrukturen von Rostsporen: Eine raster-elektronenmikroskopische Untersuchung. *Arch. Mikrobiol.* **78**, 158-165.

Brandenburger von, W., and Steiner, M. (1972). Parsitische Pilze aus Afghanistan. *Decheniana* **125**, 165-188.

Brown, A. M. (1932). Diploidization of haploid by diploid mycelium of *Puccinia helianthi* Schw. *Nature (London)* **130**, 777.

Brown, A. M. (1935). A study of coalescing haploid pustules in *Puccinia helianthi*. *Phytopathology* **25**, 1085-1090.

Brown, A. M. (1941a). Studies on the perennial rust *Puccinia minussensis*. *Can. J. Res. Sect. C.* **19**, 75-79.

Brown, A. M. (1941b). Personal communication, cited by A. H. R. Buller (1950), p. 269.

Brown, M. F., and Brotzman, H. G. (1976). Procedures for obtaining sectional views of fungal fructifications by scanning electron microscopy. *Can. J. Microbiol.* **22**, 1252-1257.

Buller, A. H. R. (1950). "Researches on Fungi," Vol. VII. University of Toronto Press, Toronto.

Buller, A. H. R. (1958). The violent discharge of aecidiospores. *In* "Researches on Fungi," Vol. III, pp. 552-559. Hafner Publ. Co., New York.

Bushnell, W. R. (1972). Physiology of fungal haustoria. *Annu. Rev. Phytopathol.* **10**, 151-176.

Bushnell, W. R. (1976). Growth of races 38 and 17, *Puccinia graminis* f. sp. *tritici*, on artificial media. *Can. J. Bot.* **54**, 1490-1498.

Calonge, F. D. (1969). Ultrastructure of the haustoria or intracellular hyphae in four different fungi. *Arch. Mikrobiol.* **67**, 209-225.

Carling, D. E., Brown, M. F., and Millikan, D. F. (1976). Ultrastructural examination of the *Puccinia graminis-Darluca filum* host-parasite relationship. *Phytopathology* **66**, 419-422.

Cassell, R. C., Woods, H. L., and Meehan, F. (1950). Spines of rust spores. *Phytopathology* **40**, 869 (Abstr.).

Chakravarti, B. P. (1966). Attempts to alter infection processes and aggressiveness of *Puccinia graminis* var. *tritici*. *Phytopathology* **56**, 223-229.

Chakravorty, A. K., and Shaw, M. (1977). A possible molecular basis for obligate host-pathogen interactions. *Biol. Rev.* **52**, 147-179.

Chrispeels, M. J. (1976). Biosynthesis, intracellular transport, and secretion of extracellular macromolecules. *Plant Physiol.* **27**, 19-38.

Christman, A. H. (1905). Sexual reproduction in the rusts. *Bot. Gaz.* **39**, 267-275.

Coffey, M. D. (1975). Obligate parasites of higher plants, particularly rust fungi. *Symp. Soc. Exp. Biol.* **29**, 297-323.

Coffey, M. D. (1976). Flax rust resistance involving the K gene: an ultrastructural survey. *Can. J. Bot.* **54**, 1443-1457.

Coffey, M. D., Palevitz, B. A., and Allen, P. J. (1972a). The fine structure of two rust fungi, *Puccinia helianthi* and *Melampsora lini*. *Can. J. Bot.* **50**, 231-240.

Coffey, M. D., Palevitz, B. A., and Allen, P. J. (1972b). Ultrastructural changes in rust-infected tissues of flax and sunflower. *Can. J. Bot.* **50**, 1485-1492.

Cole, G. T., and Aldrich, H. C. (1971). Ultrastructure of conidiogenesis in *Scopulariopsis brevicaulis. Can. J. Bot.* **49**, 745-755.

Colley, R. H. (1918). Parasitism, morphology, and cytology of *Cronartium ribicola. J. Agric. Res.* **15**, 619-660.

Corlett, M. (1970). Surface structure of the urediniospores of *Puccinia coronata* f. sp. *avenae. Can. J. Bot.* **48**, 2159-2161.

Cotter, R. V. (1960). Fertilization of pycnia with urediospores and aeciospores in *Puccinia graminis. Phytopathology* **50**, 567-568.

Couch, J. N. (1937). A new fungus intermediate between the rusts and *Septobasidium. Mycologia* **29**, 665-673.

Cox, G., and Sanders, F. (1974). Ultrastructure of the host-fungus interface in a vesicular-arbuscular mycorrhiza. *New Phytol.* **73**, 901-912.

Craigie, J. H. (1927a). Experiments in sex in rust fungi. *Nature (London)* **120**, 116-117.

Craigie, J. H. (1927b). Discovery of the function of the pycnia of the rust fungi. *Nature (London)* **120**, 765-767.

Craigie, J. H. (1933). Union of pycniospores and the haploid hyphae in *Puccinia helianthi* Schw. *Nature (London)* **131**, 25.

Craigie, J. H. (1959). Nuclear behavior in the diploidization of haploid infections of *Puccinia helianthi. Can. J. Bot.* **37**, 843-855.

Craigie, J. H., and Green, G. J. (1962). Nuclear behavior leading to conjugate association in haploid infections of *Puccinia graminis. Can. J. Bot.* **40**, 163-178.

Cummins, G. B. (1959). "Illustrated Genera of Rust Fungi." Burgess Publ. Co., Minneapolis.

Davidse, L. C., and Flach, W. (1977). Differential binding of methyl benzimidazole-2-ylcarbamate to fungal tubulin as a mechanism of resistance to this antimitotic agent in mutant strains of *Aspergillus nidulans. J. Cell Biol.* **72**, 174-193.

DeBary, A. (1887). "Comparative morphology and biology of the fungi, mycetozoa, and bacteria." Oxford University Press, Oxford.

Dickinson, S. (1949). Studies in the physiology of obligate parasitism. II. The behaviour of germ tubes of certain rusts in contact with various membranes. *Ann. Bot. (London)* **13**, 219-236.

Dickinson, S. (1955). Studies in the physiology of obligate parasitism. V. Further differences between the uredospore germ tube and leaf hyphae of *Puccinia triticina. Ann. Bot. (London)* **19**, 161-171.

Dickinson, S. (1969). Studies in the physiology of obligate parasitism. VI. Directed growth. *Phytopathol. Z.* **66**, 38-49.

Dickinson, S. (1970). Studies in the physiology of obligate parasitism. VII. The effect of a curved thigmotropic stimulus. *Phytopathol. Z.* **69**, 115-124.

Dickinson, S. (1971). Studies in the physiology of obligate parasitism. VIII. An analysis of fungal responses to thigmotropic stimuli. *Phytopathol. Z.* **70**, 62-70.

Dickinson, S. (1977). Studies in the physiology of obligate parasitism. X. Induction of responses to a thigmotropic stimulus. *Phytopathol. Z.* **89**, 97-115.

Dielman-Van Zaayen, A., Igsez, O., and Finch, J. T. (1970). Intracellular appearance and some morphological features of virus-like particles in an ascomycete fungus. *Virology* **42**, 534-537.

Dodge, B. O. (1922). Studies in the genus *Gymnosporangium*. IV. Distribution of the mycelium and the subcuticular origin of the telium in *G. clavipes. Am. J. Bot.* **9**, 354-365.

Dodge, B. O. (1924). Aecidiospore discharge as related to the character of the spore wall. *J. Agric. Res.* **27**, 749-756.

Dörr, I. (1969). Feinstruktur intrazellular wachsender *Cuscuta*-Hyphen. *Protoplasma* **67**, 123-137.

Dunkle, L. D., and Allen, P. J. (1971). Infection structure differentiation by wheat stem rust uredospores in suspension. *Phytopathology* **61**, 649-652.

Dunkle, L. D., Wergin, W. P., and Allen, P. J. (1970). Nucleoli in differentiated germ tubes of wheat stem rust uredospores. *Can. J. Bot.* **48,** 1693-1695.

Durrieu, G. (1974). Teliospore ornamentation in *Puccinia* parasitic on Umbelliferae. *Trans. Br. Mycol. Soc.* **62,** 406-411.

Dykstra, M. J. (1974). An ultrastructural examination of the structure and germination of asexual propagules of four mucoralean fungi. *Mycologia* **66,** 477-489.

Ehrlich, H. G., and Ehrlich, M. A. (1962). Fine structure of *Puccinia graminis tritici* in resistant and susceptible host varieties. *Am. J. Bot.* **49,** 665 (Abstr.).

Ehrlich, H. G., and Ehrlich, M. A. (1963). Electron microscopy of the host-parasite relationships in stem rust of wheat. *Am. J. Bot.* **50,** 123-130.

Ehrlich, M. A., and Ehrlich, H. G. (1969). Uredospore development in *Puccinia graminis. Can. J. Bot.* **47,** 2061-2064.

Ehrlich, M. A., and Ehrlich, H. G. (1970). Electron microscope radioautography of ^{14}C transfer from rust uredospores to wheat host cells. *Phytopathology* **60,** 1850-1851.

Ehrlich, M. A., and Ehrlich, H. G. (1971a). Fine structure of the host-parasite interfaces in mycoparasitism. *Annu. Rev. Phytopathol.* **9,** 155-184.

Ehrlich, M. A., and Ehrlich, H. G. (1971b). Fine structure of *Puccinia graminis* and the transfer of ^{14}C from uredospores to *Triticum vulgare. In* "Morphological and Biochemical Events in Plant-Parasite Interaction" (S. Akai and S. Ouchi, eds.), pp. 279-307. Mochizuki Publ. Co., Japan.

Ehrlich, M. A., Ehrlich, H. G., and Schafer, J. F. (1968a). Septal pores in the Heterobasidiomycetidae, *Puccinia graminis* and *P. recondita. Am. J. Bot.* **55,** 1020-1027.

Ehrlich, M. A., Schafer, J. F., and Ehrlich, H. G. (1968b). Lomasomes in wheat leaves infected by *Puccinia graminis* and *P. recondita. Can. J. Bot.* **46,** 17-20.

El-Gewely, M. R., Smith, W. E., and Colotelo, N. (1972). The reaction of near-isogenic lines of flax to the rust fungus *Melampsora lini.* I. Host-parasite interface. *Can. J. Genet. Cytol.* **14,** 743-751.

Ellingboe, A. H. (1972). Genetics and physiology of primary infection by *Erysiphe graminis. Phytopathology* **62,** 401-406.

Elnaghy, M. A., and Heitefuss, R. (1976). Permeability changes and production of antifungal compounds in *Phaseolus vulgaris* infected with *Uromyces phaseoli.* Role of the spore germination self-inhibitor. *Physiol. Plant Pathol.* **8,** 253-267.

Favali, M. A., and Marte, M. (1973). Electron microscope-autoradiography of rust-affected bean leaves labelled with tritiated glycine. *Phytopathol. Z.* **76,** 343-347.

Foster, A. S., and Gifford, E. M., Jr. (1974). "Comparative Morphology of Vascular Plants." 2nd Ed. W. H. Freeman and Co., San Francisco.

Frederick, S. A., and Newcomb, E. H. (1969). Cytochemical localization of catalase in leaf microbodies (peroxisomes). *J. Cell Biol.* **43,** 343-353.

Frederick, S. A., Newcomb, E. H., and Vigil, E. L. (1968). Fine-structural characterization of plant microbodies. *Planta (Berlin)* **81,** 229-252.

Frenster, J. H. (1974). Ultrastructure and function of heterochromatin and euchromatin. *In* "The Cell Nucleus" (H. Busch, ed.), Vol I, pp. 565-580. Academic Press, New York.

Fromme, F. D. (1914). The morphology and cytology of the aecidium cup. *Bot. Gaz.* **58,** 1-35.

Fullerton, R. A. (1970). An electron microscope study of the intracellular hyphae of some smut fungi (Ustilaginales). *Aust. J. Bot.* **18,** 285-292.

Gardner, J. S., and Hess, W. M. (1977). Ultrastructure of lipid bodies in *Tilletia caries* teliospores. *J. Bacteriol.* **131,** 622-671.

Garrett, W. N., and Wilcoxson, R. D. (1960). Fertilization of pycnia with urediospores in *Puccinia graminis* var. *tritici. Phytopathology* **50,** 636 (Abstr.).

Gay, J. L., Greenwood, A. D., and Heath, I. B. (1971). The formation and behaviour of vacuoles (vesicles) during oosphere development and zoospore germination in *Saprolegnia. J. Gen. Microbiol.* **65,** 233-241.

Gay, J. L., and Manners, J. M. (1977). Physiology of powdery mildew haustoria. *Abstr. 2nd Int. Mycol. Congr., Tampa, Florida.*

Gibson, C. M. (1904). Notes on infection experiments with various Uredineae. *New Phytol.* **3,** 184-191.

Gil, F., and Gay, J. L. (1977). Ultrastructural and physiological properties of the host interfacial components of haustoria of *Erysiphe pisi in vivo* and *in vitro. Physiol. Plant Pathol.* **10,** 1-12.

Girbardt, M. (1968). Ultrastructure and dynamics of the moving nucleus. *Symp. Soc. Exp. Biol.* **22,** 249-260.

Girbardt, M. (1971). Ultrastructure of the fungal nucleus. II. The kinetochore equivalent (KCE). *J. Cell Sci.* **10,** 453-473.

Girbardt, M., and Hädrich, H. (1975). Ultrastruktur des Pilzkernes. III. Des Kern-assoziierten Organelles (NAO = "KCE"). *Z. Allg. Mikrobiol.* **15,** 157-167.

Gold, R. E., Littlefield, L. J., and Statler, G. D. (1979). Ultrastructure of the pycnial and aecial stages of *Puccinia recondita. Can. J. Bot.* **57,** 74-86.

Gold, R. E., and Littlefield, L. J. (1979). Ultrastructure of the telial, pycnial and aecial stages of *Melampsora lini. Can. J. Bot.* **57,** (in press).

Grambow, H. J., and Riedel, S. (1977). The effect of morphologically active factors from host and nonhost plants on the *in vitro* differentiation of infection structures of *Puccinia graminis* f. sp. *tritici. Physiol. Plant Pathol.* **11,** 213-224.

Grand, L. F., and Moore, R. T. (1972). Scanning electron microscopy of *Cronartium* spores. *Can. J. Bot.* **50,** 1741-1742.

Grove, S. N., and Bracker, C. E. (1970). Protoplasmic organization of hyphal tips among fungi: Vesicles and Spitzenkörper. *J. Bacteriol.* **104,** 989-1009.

Grove, S. N., Bracker, C. E., and Morré, D. J. (1970). An ultrastructural basis for hyphal tip growth in *Pythium ultimum. Am. J. Bot.* **59,** 245-266.

Hammill, T. M. (1974). Electron microscopy of phialides and conidiogenesis in *Trichoderma saturnisporum. Am. J. Bot.* **61,** 15-24.

Hanna, W. F. (1929). Nuclear association in the aecium of *Puccinia graminis. Nature (London)* **24,** 267.

Harder, D. E. (1976a). Mitosis and cell division in some cereal rust fungi. I. Fine structure of the interphase and premitotic nuclei. *Can. J. Bot.* **54,** 981-994.

Harder, D. E. (1976b). Mitosis and cell division in some cereal rust fungi. II. The processes of mitosis and cytokinesis. *Can. J. Bot.* **54,** 995-1009.

Harder, D. E. (1976c). Electron microscopy of urediospore formation in *Puccinia coronata avenae* and *P. graminis avenae. Can. J. Bot.* **54,** 1010-1019.

Harder, D. E. (1977). Electron microscopy of teliospore formation in *Puccinia coronata avenae. Physiol. Plant Pathol.* **10,** 21-28.

Harder, D. E. (1978). Comparative ultrastructure of the haustoria in uredial and pycnial infections of *Puccinia coronata avenae. Can. J. Bot.* **56,** 214-224.

Harder, D. E., and Chong, J. (1978). Ultrastructure of spermatium ontogeny in *Puccinia coronata avenae. Can. J. Bot.* **56,** 395-403.

Harder, D. E., Rohringer, R., Samborski, D. J., Kim, W. K., and Chong, J. (1978). Electron microscopy of susceptible and resistant near isogenic (sr6/Sr6) lines of wheat infected by *Puccinia graminis tritici.* I. The host pathogen interface in the compatible (sr6/P6) interaction. *Can. J. Bot.* **56,** 2955-2966.

Hardwick, N. V., Greenwood, A. D., and Wood, R. K. S. (1971). The fine structure of the haustorium of *Uromyces appendiculatus* in *Phaseolus vulgaris. Can. J. Bot.* **49,** 383-390.

Hardwick, N. V., Greenwood, A. D., and Wood, R. K. S. (1975). Observations on the structure of uredospores of *Uromyces appendiculatus. Trans. Br. Mycol. Soc.* **64,** 289-293.

Hassan, Z. M. M., and Littlefield, L. J. (1979). Ontogeny of the uredium of *Melampsora lini*. *Can. J. Bot.* **57** (in press).

Hawker, L. E. (1965). Fine structure of fungi as revealed by electron microscopy. *Biol. Rev.* **40**, 52–92.

Heath, I. B. (1978). Experimental studies of mitosis in the fungi. *In* "Nuclear Division in the Fungi" (I. B. Heath, ed.), pp. 89–176. Academic Press, London.

Heath, I. B., Gay, J. L., and Greenwood, A. D. (1971). Cell wall formation in the Saprolegniales: Cytoplasmic vesicles underlying developing walls. *J. Gen. Microbiol.* **65**, 225–232.

Heath, I. B., and Greenwood, A. D. (1970). The structure and formation of lomasomes. *J. Gen. Microbiol.* **62**, 129–137.

Heath, I. B., and Heath, M. C. (1976). Ultrastructure of mitosis in the cowpea rust fungus *Uromyces phaseoli* var. *vignae. J. Cell Biol.* **70**, 592–607.

Heath, I. B., and Heath, M. C. (1978). Microtubules and organelle movements in the rust fungus *Uromyces phaseoli* var. *vignae. Cytobioligie* **16**, 393–411.

Heath, M. C. (1971). Haustorial sheath formation in cowpea leaves immune to rust infection. *Phytopathology* **61**, 383–388.

Heath, M. C. (1972). Ultrastructure of host and nonhost reactions to cowpea rust. *Phytopathology* **62**, 27–38.

Heath, M. C. (1974a). Light and electron microscope studies of the interactions of host and nonhost plants with cowpea rust—*Uromyces phaseoli* var. *vignae. Physiol. Plant Pathol.* **4**, 403–414.

Heath, M. C. (1974b). Chloroplast ultrastructure and ethylene production of senescing and rust-infected cowpea leaves. *Can. J. Bot.* **52**, 2591–2597.

Heath, M. C. (1975). Structure and function of fungal septa. *Proc. 1st Int. Congr. Int. Assoc. Microbiol. Soc.* **2**, 118–124.

Heath, M. C. (1976a). Hypersensitivity, the cause or consequence of rust resistance? *Phytopathology* **66**, 935–936.

Heath, M. C. (1976b). Ultrastructural and functional similarity of the haustorial neckband of rust fungi and the Casparian strip of vascular plants. *Can. J. Bot.* **54**, 2484–2489.

Heath, M. C. (1977). A comparative study of non-host interactions with rust fungi. *Physiol. Plant Pathol.* **10**, 73–88.

Heath, M. C., and Heath, I. B. (1971). Ultrastructure of an immune and a susceptible reaction of cowpea leaves to rust infection. *Physiol. Plant Pathol.* **1**, 277–287.

Heath, M. C., and Heath, I. B. (1975). Ultrastructural changes associated with the haustorial mother cell during haustorium formation in *Uromyces phaseoli* var. *vignae. Protoplasma* **84**, 297–314.

Heath, M. C., and Heath, I. B. (1978). Structural studies of the development of infection structures of cowpea rust, *Uromyces phaseoli* var. *vignae.* I. Nucleoli and nuclei. *Can. J. Bot.* **56**, 648–661.

Henderson, D. M. (1969). Studies in the morphology of fungal spores. I: The teliospores of *Puccinia prostii* and *Nyssopsora echinata. Notes R. Bot. Gard. Edinburgh* **29**, 373–375.

Henderson, D. M. (1973). Studies in the morphology of fungal spores: *Trachyspora intrusa. Rep. Tottori Mycol. Inst.* **10**, 163–168.

Henderson, D. M., and Hiratsuka, Y. (1974). Ontogeny of spore markings on aeciospores of *Cronartium comandrae* and peridermioid teliospores of *Endocronartium harknessii. Can. J. Bot.* **52**, 1919–1921.

Henderson, D. M., and Prentice, H. T. (1973). Development of the spores of *Phragmidium. Nova Hedwigia Z. Kryptogamenkd.* **24**, 431–441.

Henderson, D. M., and Prentice, H. T. (1974). Spore morphogenesis of *Coleosporium tussilaginus. Trans. Br. Mycol. Soc.* **63**, 431–435.

Henderson, D. M., Eudall, R., and Prentice, H. T. (1972a). Morphology of the reticulate telio-spores of *Puccinia chaerophylli*. *Trans. Br. Mycol. Soc.* **59**, 229-232.

Henderson, D. M., Prentice, H. T., and Eudall, R. (1972b). The morphology of fungal spores: III. The aeciodiospores of *Puccinia poarum*. *Pollen Spores* **14**, 17-24.

Hennen, J. F., and Figueiredo, M. B. (1979). *Intrapes*, a new genus of Fungi Imperfecti (Ure-dinales) from the Brazilian cerrado. *Mycologia* (submitted).

Hersperger, C. (1928). Ueber das Vorkommen einer Aecidienperidie bei Melampsoren. *Naturforsch. Geselsch.*, Bern 5, XXVII.

Heslop-Harrison, J., (1966). Cytoplasmic continuities during spore formation in flowering plants. *Endeavour* **25**, 65-72.

Hess, S. L., Allen, P. J., and Lester, H. (1976). Uredospore wall digestion during germination independent of macromolecular synthesis. *Physiol. Plant Pathol.* **9**, 265-272.

Hess, S. L., Allen, P. J., Nelson, D., and Lester, H. (1975). Mode of action of methyl *cis*-ferulate, the self-inhibitor of stem rust uredospore germination. *Physiol. Plant Pathol.* **5**, 107-112.

Hess, W. M., and Weber, D. J. (1976). Form and function of Basidiomycete spores. *In* "The Fungal Spore, Form and Function" (D. J. Weber and W. M. Hess, eds.), pp. 645-714. John Wiley and Sons, New York.

Hickey, E. L., and Coffey, M. D. (1977). A fine-structural study of the pea downy mildew fungus *Peronospora pisi* in its host *Pisum sativum*. *Can. J. Bot.* **55**, 2845-2858.

Hilu, H. M. (1965). Host-pathogen relationships of *Puccinia sorghi* in nearly isogenic resistant and susceptible seedling corn. *Phytopathology* **55**, 563-569.

Hiratsuka, N., and Hiratsuka, Y. (1977). Morphology of spermogonia and taxonomy of rust fungi. *Proc. 2nd Int. Mycol. Congr.* **1**, 288 (Abstr.)

Hiratsuka, N., and Kaneko, S. (1975). Surface structure of *Coleosporium* spores. *Rept. Tottori Mycol. Inst.* **12**, 1-13.

Hiratsuka, Y. (1965). The identification of *Uraecium holwayi* on hemlock as the aecial state of *Pucciniastrum vaccinii* in western North America. *Can. J. Bot.* **43**, 475-478.

Hiratsuka, Y. (1968). Morphology and cytology of aeciospores and aeciospore germ tubes of host-alternating and pine-to-pine races of *Cronartium flaccidium* in northern Europe. *Can. J. Bot.* **46**, 1119-1122.

Hiratsuka, Y. (1969). *Endocronartium*, a new genus for autoecious pine rusts. *Can. J. Bot.* **47**, 1493-1495.

Hiratsuka, Y. (1970). Emergence of the aeciospore germ tubes of *Cronartium coleosporioides* (= *Peridermium stalactiforme*) as observed by scanning electron microscope. *Can. J. Bot.* **48**, 1692.

Hiratsuka, Y. (1971). Spore surface morphology of pine stem rusts of Canada as observed under a scanning electron microscope. *Can. J. Bot.* **49**, 371-372.

Hiratsuka, Y. (1973). The nuclear cycle and the terminology of spore states in Uredinales. *Mycologia* **65**, 432-443.

Hiratsuka, Y., and Cummins, G. B. (1963). Morphology of the spermogonia of the rust fungi. *Mycologia* **55**, 487-507.

Hiratsuka, Y., and Powell, J. M. (1976). Pine stem rusts of Canada. *For. Tech. Rept. 4, Can. For. Serv. Ottawa.*

Hoch, H. C. (1977). Mycoparasitic relationships. III. Parasitism of *Physalospora obtusa* by *Calcarisporium parasiticum*. *Can. J. Bot.* **55**, 198-207.

Hofsten von, A., and Holm, L. (1968). Studies on the fine structure of aeciospores. I. *Grana Palynol.* **8**, 235-251.

Holm, L., Dunbar, A., and Hofsten von, A. (1970). Studies on the fine structure of aeciospores. II. *Sven. Bot. Tidskr.* **64**, 380-382.

Holm, L., and Tibell, L. (1974). Studies on the fine structure of aeciospores. III. Aeciospore ontogeny in *Puccinia graminis*. *Sven. Bot. Tidskr.* **68**, 136-152.

Hoppe, H. H., and Heitefuss, R. (1974). Permeability and membrane lipid metabolism of *Phaseolus vulgaris* infected with *Uromyces phaseoli*. I. Changes in the efflux of cell constituents. *Physiol. Plant Pathol.* **4**, 5-9.

Hughes, S. J. (1970). Ontogeny of spore forms in Uredinales. *Can. J. Bot.* **48**, 2147-2157.

Hung, C. Y., and Wells, K. (1977). The behavior of the nucleolus during nuclear divisions in the asci of *Pyronema domesticum*. *Mycologia* **69**, 685-692.

Hunt, P. (1968). Cuticular penetration by germinating uredospores. *Trans. Br. Mycol. Soc.* **51**, 103-112.

Huttinga, H., Wichers, H. J., and Dielman-Van Zaayen, A. (1975). Filamentous and polyhedral virus-like particles in *Boletus edulis*. *Neth. J. Plant Pathol.* **81**, 102-106.

Johnson, T. (1934). A tropic response in germ tubes of urediospores of *Puccinia graminis tritici*. *Phytopathology* **24**, 80-82.

Jones, D. R. (1971). Surface structure of germinating *Uromyces dianthi* urediospores as observed by scanning electron microscope. *Can. J. Bot.* **49**, 2243.

Jones, D. R. (1973). Ultrastructure of septal pore in *Uromyces dianthi*. *Trans. Br. Mycol. Soc.* **61**, 227-235.

Jones, D. R., and Deverall, B. J. (1977). The effect of the Lr20 resistance gene in wheat on the development of leaf rust, *Puccinia recondita*. *Physiol. Plant Pathol.* **10**, 275-284.

Kajiwara, T. (1971). Structure and physiology of haustoria of various parasites. In "Morphological and Biochemical Events in Plant-Parasite Interaction" (S. Akai and S. Ouchi, eds.), pp. 255-277. Mochizuki Publishing Co., Japan.

Kajiwara, T. and Takahashi, K. (1975). Ultrastructural changes in host-parasite interface of chrysanthemum rust treated with oxycarboxin. *Proc. 1st Int. Cong. Int. Assoc. Microbiol. Soc.* **2**, 131-135.

Kenney, M. J. (1970). Comparative morphology of the uredia of the rust fungi. Doctoral Thesis, Purdue University, Lafayette, Indiana.

Kern, F. D. (1910). The morphology of the peridial cells in the roestelioid. *Bot. Gaz.* **49**, 445-451.

Kern, F. D. (1973). "A Revised Taxonomic Account of *Gymnosporangium*." The Pennsylvania State University Press, University Park.

Kim, W. K., Rohringer, R., Samborski, D. J., and Howes, N. K. (1977). Effect of blasticidin S, ethionine, and polyoxin D on stem rust development and host-cell necrosis in wheat near-isogenic for gene Sr6. *Can. J. Bot.* **55**, 568-573.

Kinden, D. A., and Brown, M. F. (1975a). Technique for scanning electron microscopy of fungal structures within plant cells. *Phytopathology* **65**, 74-76.

Kinden, D. A., and Brown, M. F. (1975b). Electron microscopy of vesicular-arbuscular mycorrhizae of yellow poplar. II. Intracellular hyphae and vesicles. *Can. J. Microbiol.* **21**, 1768-1780.

Kiraly, Z., Barna, B., and Ersek, T. (1972). Hypersensitivity as a consequence, not the cause, of plant resistance to infection. *Nature (London)* **239**, 456-458.

Kohno, M., Nishimura, T., Ishizaki, H., and Kunoh, H. (1975a). Cytological studies on rust fungi. II. Ultrastructure of sporidia of *Puccinia horiana* P. Hennings. *Bull. Fac. Agric. Mie Univ. (Japan)* **48**, 9-15.

Kohno, M., Nishimura, T., Ishizaki, H., and Kunoh, H. (1975b). Ultrastructure changes of cell wall in germinating teliospore of *Gymnosporangium haraeanum* Sydow. *Trans. Mycol. Soc. Jpn.* **16**, 106-112.

Kohno, M., Ishizaki, H., and Kunoh, H. (1976). Cytological studies on rust fungi. V. Intracellular hyphae of *Gymnosporangium haraeanum* Sydow in cells of Japanese pear leaves. *Ann. Phytopathol. Soc. Jpn.* **42**, 417-423.

Kohno, M., Ishizaki, H., and Kunoh, H. (1977a). Cytological studies on rust fungi. VI. Fine structures of infection process of *Kuehneola japonica* (Diet.) Dietel. *Mycopathologia* **61**, 35-42.

Kohno, M., Nishimura, T., Noda, M., Ishizaki, H., and Kunoh, H. (1977b). Cytological studies on rust fungi. VII. The nuclear behavior of *Gymnosporangium asiaticum* Miyabe et Yamada during the stages from teliospore germination through sporidium germination. *Trans. Mycol. Soc. Jpn.* **18**, 211–219.

Kozar, F., and Netolitzky, H. J. (1975). Ultrastructure and cytology of pycnia, aecia, and aeciospores of *Gymnosporangium clavipes*. *Can. J. Bot.* **53**, 972–977.

Kunoh, H., Ishizaki, H., and Kohno, M. (1977). Application of the scanning electron microscope to plant pathology. *Fitopathol. Brasil.* **2**, 131–153.

Lalonde, M., and Knowles, R. (1975). Ultrastructure, composition, and biogenesis of the encapsulation material surrounding the endophyte in *Alnus crispa* var. *mollis* root nodules. *Can. J. Bot.* **53**, 1951–1971.

Lamb, I. M. (1935). The initiation of the dikaryophase in *Puccinia phragmitis* (Schum.) Körn. *Ann. Bot. (London)* **49**, 403–438.

Laundon, G. F. (1973). Uredinales. In "The Fungi, An Advanced Treatise, IVB" (C. G. Ainsworth, F. K. Sparrow, and A. S. Sussman, eds.), pp. 247–279. Academic Press, New York.

Leath, K. T., and Rowell, J. B. (1966). Histological study of the resistance of *Zea mays* to *Puccinia graminis*. *Phytopathology* **56**, 1305–1309.

Leath, K. T., and Rowell, J. B. (1969). Thickening of corn mesophyll cell walls in response to invasion by *Puccinia graminis*. *Phytopathology* **59**, 1654–1656.

Lecog, H., Spire, D., Rapilly, F., and Bertrandy, J. (1974). Mise en évidence de particules de type viral chez les *Puccinia*. *C. R. Hebd. Seances Acad. Sci., Ser. D* **279**, 1599–1602.

Lemke, P. A. (1976). Viruses of eukaryotic microorganisms. *Annu. Rev. Microbiol.* **30**, 29–56.

Leppik, E. E. (1956). Some viewpoints on the phylogeny of rust fungi. II. *Gymnosporangium fuscum*. *Mycologia* **48**, 637–654.

Leppik, E. E. (1977). Form and function of balanoid aecia of *Gymnosporangium fuscum*. *Mycologia* **69**, 967–974.

Lewis, B. G., and Day, J. R. (1972). Behaviour of uredospore germ-tubes of *Puccinia graminis tritici* in relation to the fine structure of wheat leaf surfaces. *Trans. Br. Mycol. Soc.* **58**, 139–145.

Littlefield, L. J. (1971a). Ultrastructure of septa in Uredinales. 1st Int. Mycol. Cong. Exeter p. 58 (Abstr.).

Littlefield, L. J. (1917b). Scanning electron microscopy of urediospores of *Melampsora lini*. *J. Microsc. (Paris)* **10**, 225–228.

Littlefield, L. J. (1972). Development of haustoria of *Melampsora lini*. *Can. J. Bot.* **50**, 1701–1703.

Littlefield, L. J. (1973). Histological evidence for diverse mechanisms of resistance to flax rust, *Melampsora lini* (Ehrenb.) Lev. *Physiol. Plant Pathol.* **3**, 241–247.

Littlefield, L. J. (1974). Scanning electron microscopy of internal tissues of rust-infected flax. *Trans. Br. Mycol. Soc.* **63**, 208–211.

Littlefield, L. J., and Aronson, S. J. (1969). Histological studies of *Melampsora lini* resistance in flax. *Can. J. Bot.* **47**, 1713–1717.

Littlefield, L. J., and Bracker, C. E. (1970). Continuity of host plasma membrane around haustoria of *Melampsora lini*. *Mycologia* **62**, 609–614.

Littlefield, L. J., and Bracker, C. E. (1971a). Ultrastructure of septa in *Melampsora lini*. *Trans. Br. Mycol. Soc.* **56**, 181–188.

Littlefield, L. J., and Bracker, C. E. (1971b). Ultrastructure and development of urediospore ornamentation in *Melampsora lini*. *Can. J. Bot.* **49**, 2067–2073.

Littlefield L. J., and Bracker, C. E. (1972). Ultrastructural specialization of the host-pathogen interface in rust-infected flax. *Protoplasma* **74**, 271–305.

Locci, R., and Bisiach, M. (1970). The *Phaseolus vulgaris–Uromyces appendiculatus* complex.

I. Examination of the uredospore infection process by scanning electron microscopy. *Riv. Patal. Veg.* **6,** 21–28.

Longo, N., and Naldini, B. (1970). Osservazioni sui pori dei setti in *Melampsora larici-tremulae* Klebahn e *Melampsora pinitorqua* (Braun) Rostr. *Caryologia* **23,** 657–672.

Lovett, J. S. (1976). Regulation of protein metabolism during spore germination. *In* "The Fungal Spore, Form and Function" (D. J. Weber and W. M. Hess, eds.), pp. 182–242. John Wiley and Sons, New York.

Lupton, F. G. H. (1956). Resistance mechanisms of species of *Triticum* and *Aegilops* and of amphidiploids between them to *Erysiphe graminis* D. C. *Trans. Br. Mycol. Soc.* **39,** 51–59.

Macko, V., Staples, R. C., Yaniv, A., and Granados, R. R. (1976). Self-inhibitors of fungal spore germination. *In* "The Fungal Spore, Form and Function" (D. J. Weber and W. M. Hess, eds.), pp. 73–100. John Wiley and Sons, New York.

Maheshwari, R. (1966). The physiology of penetration and infection by urediospores of rust fungi. Doctoral Thesis, University of Wisconsin, Madison.

Maheshwari, R., Allen, P. J., and Hildebrandt, A. C. (1967). Physical and chemical factors controlling the development of infection structures from urediospore germ tubes of rust fungi. *Phytopathology* **57,** 855–862.

Maheshwari, R., and Hildebrandt, A. C. (1967). Directional growth of the urediospore germ tubes and stomatal penetration. *Nature (London)* **214,** 1145–1146.

Maheshwari, R., Hildebrandt, A. C., and Allen, P. J. (1965). Differentiation of infection structure in uredospore germ tubes. *Am. J. Bot.* **52,** 632 (Abstr.).

Manners, J. M., and Gay, J. L. (1978). Uptake of ^{14}C photosynthates from *Pisum sativum* by haustoria of *Erysiphe pisi. Physiol. Plant Pathol.* **11,** 199–209.

Manocha, M. S. (1971). Presence of nucleolus in flax rust axenic culture. *Phytopathol. Z.* **71,** 113–118.

Manocha, M. S. (1975). Autoradiography and fine structure of host–parasite interface in temperature-sensitive combinations of wheat stem rust. *Phytopathol. Z.* **82,** 207–215.

Manocha, M. S., and Lee, K. Y. (1972). Host–parasite relations in mycoparasite. II. Incorporation of tritiated *N*-acetylglucosamine into *Choanephora cucurbitarum* infected with *Piptocephalis virginiana. Can. J. Bot.* **50,** 35–37.

Manocha, M. S., and Letourneau, D. R. (1978). Structure and composition of host–parasite interface in a mycoparasite system. *Physiol. Plant Pathol.* **12,** 141–150.

Manocha, M. S., and Shaw, M. (1966). The physiology of host–parasite relations. XVI. Fine structure of the nucleus in the rust-infected mesophyll cells of wheat. *Can. J. Bot.,* **44,** 669–673.

Manocha, M. S., and Shaw, M. (1967). Electron microscopy of uredospores of *Melampsora lini* and of rust-infected flax. *Can. J. Bot.* **45,** 1575–1582.

Manocha, M. S., and Wisdom, C. J. (1971). Ultrastructural studies on the uredial stage of wheat stem rust. *Phytopathol. Z.* **70,** 263–273.

Marchetti, M. A., Vecker, F. A., and Bromfield, K. R. (1975). Uredial development of *Phakopsora pachyrhizi* in soybeans. *Phytopathology* **65,** 822–823.

Martin, G. W. (1957). The Tulasnelloid fungi and their bearing on basidial terminology. *Brittonia* **9,** 25–30.

Maxwell, D. P., Armentrout, V. N., and Graves, L. B., Jr. (1977). Microbodies in plant pathogenic fungi. *Annu. Rev. Phytopathol.* **15,** 119–134.

Mayama, S., Rehfeld, D. W., and Daly, J. M. (1975). A comparison of the development of *Puccinia graminis tritici* in resistant and susceptible wheat based on glucosamine content. *Physiol. Plant Pathol.* **7,** 243–257.

McCully, E. K., and Robinow, C. F. (1972a). Mitosis in Heterobasidiomycetous yeasts. I. *Leucosporidium scottii (Candida scottii). J. Cell Sci.* **10,** 857–882.

McCully, E. K., and Robinow, C. F. (1972b). Mitosis in Heterobasidiomycetous yeasts. II.

Rhodosporidium sp. (*Rhodotorula glutinis*) and *Aessosporon salmonicolor* (*Sporobolomyces salmonicolor*). *J. Cell Sci.* **11**, 1–32.

McDonald, J. G., and Heath, M. C. (1978). Rod-shaped and spherical viruslike particles in cowpea rust fungus. *Can. J. Bot.* **56**, 963–975.

Melander, L. W., and Craigie, J. H. (1927). Nature of resistance of *Berberis* spp. to *Puccinia graminis. Phytopathology* **17**, 95–114.

Mendgen, K. (1973a). Feinbau der Infektionsstrukturen von *Uromyces phaseoli. Phytopathol. Z.* **78**, 109–120.

Mendgen, K. (1973b). Microbodies (glyoxysomes) in infection structures of *Uromyces phaseoli. Protoplasma* **78**, 477–482.

Mendgen, K. (1975). Ultrastructural demonstration of different peroxidase activities during the bean rust infection process. *Physiol. Plant Pathol.* **6**, 275–282.

Mendgen, K. (1977). Reduced lysine uptake by bean rust haustoria in a resistant reaction. *Naturwissenschaften* **64**, 438.

Mendgen, K. (1978). Der Infektionsverlauf von *Uromyces phaseoli* bei anfälligen und resistenten Bohnensorten. *Phytopathol. Z.* **93**, 259–313.

Mendgen, K., and Fuchs, W. H. (1973). Elektronenmikroskopische darstellung peroxydatischer aktivitaten bei *Phaseolus vulgaris* nach infektion mit *Uromyces phaseoli typica. Arch. Mikrobiol.* **88**, 181–192.

Mendgen, K., and Heitefuss, R. (1975). Micro-autoradiographic studies on host–parasite interactions. I. The infection of *Phaseolus vulgaris* with tritium labeled uredospores of *Uromyces phaseoli. Arch. Microbiol.* **105**, 193–199.

Mielke, J. L., and Cochran, G. W. (1952). Differences in spore surface markings of three pine rusts, as shown by the electron microscope. *Mycologia* **44**, 325–329.

Mims, C. W. (1977a). Fine structure of basidiospores of the cedar-apple rust fungus *Gymnosporangium juniperi-virginianae. Can. J. Bot.* **55**, 1057–1063.

Mims, C. W. (1977b). Ultrastructure of teliospore formation in the cedar-apple rust fungus *Gymnosporangium juniperi-virginianae. Can. J. Bot.* **55**, 2319–2329.

Mims, C. W., and Glidewell, D. C. (1978). Some ultrastructural observations on the host-pathogen relationship within the telial gall of the rust fungus *Gymnosporangium juniperi-virginianae. Bot. Gaz.* **139**, 11–17.

Mims, C. W., Seabury, F., and Thurston, E. L. (1975). Fine structure of teliospores of the cedar-apple rust *Gymnosporangium juniperi-virginianae. Can. J. Bot.* **53**, 544–552.

Mims, C. W., Seabury, F., and Thurston, E. L. (1976). An ultrastructural study of spermatium formation in the rust fungus *Gymnosporangium juniperi-virginianae. Am. J. Bot.* **63**, 997–1002.

Mitchell, S., and Shaw, M. (1969). The nucleolus of wheat stem rust uredospores. *Can. J. Bot.* **47**, 1887–1889.

Mlodzianowski, F., and Siwecki, R. (1975). Ultrastructural changes in chloroplasts of *Populus tremula* L. leaves affected by the fungus *Melampsora pinitorqua* Braun Rostr. *Physiol. Plant Pathol.* **6**, 1–4.

Moore, R. T. (1963a). Fine structure of mycota. X. Thallus formation in *Puccinia podophylli* aecia. *Mycologia* **55**, 633–642.

Moore, R. T. (1963b). Fine structure of mycota. XI. Occurrence of the Golgi dictyosome in the Heterobasidiomycete *Puccinia podophylli. J. Bacteriol.* **86**, 866–871.

Moore, R. T. (1965). The ultrastructure of fungal cells. *In* "The Fungi," (G. C. Ainsworth and A. S. Sussman, eds.), Vol. 1, pp. 95–118. Academic Press, New York.

Moore, R. T., and Grand, L. F. (1970). Application of scanning electron microscopy to basidiomycete taxonomy. *Proc. 3rd Annu. Scanning Electron Microsc. Symp. ITT Res. Inst.*, 137–144.

Moore, R. T., and McAlear, J. H. (1961). Fine structure of mycota. 8. On the aecidial stage of *Uromyces caladii. Phytopathol. Z.* **42**, 297–304.

Morré, D. J. (1975). Membrane biogenesis. *Plant Physiol.* **26**, 441-481.

Mooré, D. J., Mollenhauer, H. H., and Bracker, C. E. (1971). Origin and continuity of Golgi apparatus. *In* "Results and Problems in Cell Differentiation. Vol. 2, Origin and Continuity of Cell Organelles" (J. Reinert and H. Ursprung, eds.), pp. 82-126. Springer-Verlag, Berlin.

Müller, L. Y., Rijkenberg, F. H. J., and Truter, S. J. (1974a). Ultrastructure of the uredial stage of *Uromyces appendiculatus*. *Phytophylactica* **6**, 73-104.

Müller, L. Y., Rijkenberg, F. H. J., and Truter, S. J. (1974b). A preliminary ultrastructural study on the *Uromyces appendiculatus* teliospore stage. *Phytophylactica* **6**, 123-128.

Müller, L. Y., Rijkenberg, F. H. J., and Truter, S. J. (1974c). Light and electron microscopy of sporophore elongation in uredia of *Uromyces appendiculatus*. *Phytophylactica* **6**, 129-130.

Murrill, W. A. (1905). Terms applied to the surface and surface appendages of fungi. *Torreya* **5**, 60-66.

Mussell, H. W., Wood, H. A., Adler, J. P., and Bozarth, R. F. (1973). *Puccinia graminis:* virus-like particles in mycelium from axenic cultures. *2nd Cong. Plant Pathol. Minneapolis.* Abstr. No. 0908.

Newton, M., Johnson, T., and Brown, A. M. (1930). A preliminary study on the hybridization of physiologic forms of *Puccinia graminis tritici*. *Sci. Agric.* **10**, 728-731.

O'Donnell, K. L., and Hooper, G. R. (1977). Cryofracturing as a technique for the study of fungal structure in the scanning electron microscope. *Mycologia* **69**, 309-320.

Oláh, G. M., Pinon, J., and Janitor, A. (1971). Modifications ultrastructurales provoquées par *Puccinia gramanis* f. sp. *tritici* chez le blé. *Mycopath. Mycol. Appl.* **44**, 325-346.

Olive, L. S. (1949). Karyogamy and meiosis in the rust *Coleosporium vernoniae*. *Am. J. Bot.* **36**, 41-54.

Onoe, T., Tani, T., and Naito, N. (1972). Scanning electron microscopy of crown rust appressorium produced on oat leaf surface. *Tech. Bull. Fac. Agric. Kagawa Univ. (Japan)* **24**, 42-47.

Onoe, T., Tani, T., and Naito, N. (1973). The uptake of labeled nucleosides by *Puccinia coronata* grown on susceptible leaves. *Rep. Tottori Mycol. Inst.* **10**, 303-312.

Onoe, T., Tani, T., and Naito, N. (1976). Changes in fine structure of Shokan 1 oat leaves at the early stage of infection with incompatible race 226 of crown rust fungus. *Ann. Phytopathol. Soc. Jpn.* **42**, 481-488.

Orcival, J. (1969). Infrastructure des sucoirs et relations hôte-parasite dans des stades écidiens d'Uredinales. *C. R. Hebd. Seances Acad. Sci., Ser. D,* **269**, 1973-1975.

Orcival, J. (1972). Structure et évolution des chloroplasts dans la feuille d'*Euphorbia amygdaloides* parasitée par *Endophyllum euphorbiae-sylvaticae*. *Ann. Sci. Nat. Bot. Paris* **13**, 232-346.

Pady, S. M. (1935a). Aeciospore infection in *Gymnoconia interstitialis* by penetration of the cuticle. *Phytopathology* **25**, 453-474.

Pady, S. M. (1935b). The role of intracellular mycelium in systemic infections of *Rubus* with the orange-rust. *Mycologia* **27**, 618-637.

Pady, S. M., Kramer, C. L., and Clary, R. (1968). Periodicity in aeciospore release in *Gymnosporangium juniperi-virginianae*. *Phytopathology* **58**, 329-331.

Paliwal, Y. C., and Kim, W. K. (1974). Scanning electron microscopy of differentiating and non-differentiating uredosporelings of wheat stem rust fungus (*Puccinia graminis* f. sp. *tritici*) on an articificial substance. *Tissue & Cell* **6**, 391-397.

Parmelee, J. A. (1965). The genus Gymnosporangium in eastern Canada. *Can. J. Bot.* **43**, 239-267.

Patton, R. F., and Johnson, D. W. (1970). Mode of penetration of needles of eastern white pine by *Cronartium ribicola*. *Phytopathology* **60**, 977-982.

Patton, R. F., and Spear, R. N. (1977). Stomatal influences on infection by *Cronartium ribicola. Proc. Am. Phytopathol. Soc.* **4**, 85. (Abstr.).

Pavgi, M. S., and Flangas, A. L. (1965). Comparative electron microscopic observations on the spore forms of three *Puccinia* species. *Mycopath. Mycol. Appl.* **26**, 104-110.

Payak, M. M. (1956). A study of the pycnia, flexuous hyphae, and nuclear migrations in the aecia of *Scopella gentilis. Bot. Gaz.* **118**, 37-42.

Payak, M. M., Joshi, L. M., and Mehta, S. C. (1967). Electron microscopy of urediospores of *Puccinia graminis* var. *tritici. Phytopathol. Z.* **60**, 196-199.

Petersen, R. H. (1974). The rust fungus life cycle. *Bot. Rev.* **40**, 453-513.

Pierson, R. K. (1933). Fusion of pycniospores with filamentous hyphae in the pycnium of the white pine blister rust. *Nature (London)* **131**, 728-729.

Pinon, J. D., Bossányi, G., and Oláh, G. M. (1972). Changements ultrastructuraux observés dans les cellules de blé infecté par *Puccinia graminis* f. sp. *tritici. Acta Phytopath. Acad. Sci. Hung.* **7**, 21-39.

Plotnikova, J. M., Andreev, L. N., and Sassen, M. M. A. (1977). Host-pathogen interface of *Puccinia striiformis* haustoria and wheat cells. *2nd Int. Mycol. Cong. Tampa, Florida* **2**, 525.

Plotnikova, Y. M., Littlefield, L. J., and Miller, J. D. (1979). Scanning electron microscopy of the haustorium-host interface regions in *Puccinia graminis tritici* infected wheat. *Physiol. Plant Pathol.* **13** (in press).

Pollard, T. D. (1976). Cytoskeletal functions of cytoplasmic contractile proteins. *J. Supramol. Struct.* **5**, 317-334.

Poon, N. H., and Day, A. W. (1976). Somatic nuclear division in the sporidia of *Ustilago violacea.* IV. Microtubules and the spindle-pole body. *Can. J. Microbiol.* **22**, 507-522.

Pring, R. J. (1975) Scanning electron microscopy of rust haustoria. *J. Microsc. (Oxford)* **103**, 289-291.

Pring, R. J., and Richmond, D. V. (1975). Three-dimensional morphology of rust haustoria. *Trans. Br. Mycol. Soc.* **65**, 291-294.

Pring, R. J., and Richmond, D. V. (1976). An ultrastructural study of the effect of oxycarboxin on *Uromyces phaseoli* infecting leaves of *Phaseolus vulgaris. Physiol. Plant Pathol.* **8**, 155-162.

Punithalingam, E., and Jones, D. (1971). Aecidium species on *Agathis. Trans. Br. Mycol. Soc.* **57**, 325-331.

Rajendren, R. B. (1970). Cytology and developmental morphology of *Kernkampella breyniae-patentis* and *Ravenelia hobsoni. Mycologia* **62**, 1112-1121.

Rajendren, R. B. (1972). Evolution of haustoria in tropical rust fungi. *Bull. Torrey Bot. Club* **99**, 84-88.

Rambourg, M. A. (1967). Détection des glycoproteines en microscopie électronique: coloration de la surface cellulaire et de l'appareil de Golgi par un mélange acide chromique-phosphotungstique. *C. R. Hebd. Seances Acad. Sci., Ser. D* **265**, 1426-1428.

Rapilly, F. (1968). Quelques remarques sur la morphologie des uredospores de *Puccinia striiformis* f. sp. *tritici. Bull. Soc. Mycol. Fr.* **84**, 493-496.

Rawlinson, C. J., and MacLean, D. J. (1973). Virus-like particles in axenic cultures of *Puccinia graminis tritici. Trans. Br. Mycol. Soc.* **61**, 590-593.

Reisener, H. J. (1976). Lipid metabolism of fungal spores during sporogenesis and germination. *In* "The Fungal Spore, Form and Function" (D. J. Weber and W. M. Hess, eds.), pp. 165-188. John Wiley and Sons, New York.

Reynolds, B. A. (1975). Cytology and ultrastructure of brown rust in barley cultivars of varied resistance. Doctoral Thesis, University of London.

Rice, M. A. (1927). The haustoria of certain rusts and the relation between host and pathogene. *Bull. Torrey Bot. Club* **54**, 63-153.

Richmond, D. V. (1975). Effects of toxicants on the morphology and fine structure of fungi. *Adv. Appl. Microbiol.* **19**, 289-319.

Richmond, D. V., and Pring, R. J. (1971). The effect of benomyl on the structure of *Botrytis fabae*. *J. Gen. Microbiol.* **66**, 79-94.

Rijkenberg, F. H. J. (1972). Fine structure of the poplar rust (*Melampsora larici-populina*). *Phytophylactica* **4**, 33-40.

Rijkenberg, F. H. J. (1975). The uredial stage of maize rust. *Proc. Electron Microsc. Soc. South. Afr.* **5**, 35-36.

Rijkenberg, F. H. J., and Müller, L. Y. (1971). An ultrastructural comparison between inter- and intracellular mycelium and haustoria of rust fungi. *Proc. Electron Microsc. Soc. South. Afr.* **1**, 25-26.

Rijkenberg, F. H. J., and Truter, S. J. (1973a). Haustoria and intracellular hyphae in the rusts. *Phytopathology* **63**, 281-286.

Rijkenberg, F. H. J., and Truter, S. J. (1973b). Sporogenesis of *Puccinia sorghi* on the alternate host. *Proc. Electron Microsc. Soc. South. Afr.* **3**, 3-4.

Rijkenberg, F. H. J., and Truter, S. J. (1974a). The ultrastructure of sporogenesis in the pycnial stage of *Puccinia sorghi*. *Mycologia* **66**, 319-326.

Rijkenberg, F. H. J., and Truter, S. J. (1974b). The ultrastructure of the *Puccinia sorghi* aecial stage. *Protoplasma* **81**, 231-245.

Rijkenberg, F. H. J., and Truter, S. J. (1975). Cell fusion in the aecium of *Puccinia sorghi*. *Protoplasma* **83**, 233-246.

Rijo, L., and Sargent, J. A. (1974). The fine structure of the coffee leaf rust, *Hemileia vastatrix*. *Can. J. Bot.* **52**, 1363-1367.

Robb, J., Harvey, A. E., and Shaw, M. (1973). Ultrastructure of the hyphal walls and septa of *Cronartium ribicola* on tissue cultures of *Pinus monticola*. *Can. J. Bot.* **51**, 2301-2305.

Robb, J., Harvey, A. E., and Shaw, M. (1975a). Ultrastructure of tissue cultures of *Pinus monticola* infected by *Cronartium ribicola*. I. Prepenetration host changes. *Physiol. Plant Pathol.* **5**, 1-8.

Robb, J., Harvey, A. E., and Shaw, M. (1975b). Ultrastructure of tissue cultures of *Pinus monticola* infected by *Cronartium ribicola*. II. Penetration and post-penetration. *Physiol. Plant Pathol.* **5**, 9-18.

Roland, M. J. C. (1969). Mise en évidence sur coups ultrafins de formations polysaccharidiques directement associées au plasmalemme. *C. R. Hebd. Seances Acad. Sci., Ser. D* **269**, 939-942.

Roncadori, R. W. (1968). The pathogenicity of secondary and tertiary basidiospores of *Cronartium fusiforme*. *Phytopathology* **58**, 712-713.

Rossetti, V., and Morel, G. (1958). Le développement du *Puccinia antirrhini* sur tissue du Muflier cultives *in vitro*. *C. R. Hebd. Seances Acad. Sci., Ser. D* **247**, 1893-1895.

Rothman, P. G. (1960). Host-parasite interactions of eight varieties of oats infected with race 202 of *Puccinia coronata* var. *avenae*. *Phytopathology* **50**, 914-918.

Sappin-Trouffy, P. (1896). Sur la signification de la fecondation chez les Uredinees. *Le Botaniste* **5**, 33-244.

Savile, D. B. O. (1939). Nuclear structure and behavior in species of the Uredinales. *Am. J. Bot.* **26**, 585-609.

Savile, D. B. O. (1976). Evolution of the rust fungi (Uredinales) as reflected by their ecological problems. *In* "Evolutionary Biology" (M. K. Hecht, W. C. Steere, and B. Wallace, eds.), vol. 9, pp. 137-207. Plenum Publishing Co., New York.

Schroeder von, H. and Hassebrauk, K. (1957). Beitrage zur Biologie von *Darluca filum* (Biv.) Cast. und einigen anderen auf Uredineen beobachten Pilzen. *Zentralbl. Bakteriol. Parasitenkd. Infektionskr. Hyg. Abt. 2*, **110**, 676-696.

Shaw, M. (1967). Cell biological aspects of host-parasite relations of obligate fungal parasites. *Can. J. Bot.* **45**, 1205-1220.

Shaw, M., and Manocha, M. S. (1965a). Fine structure in detached senescing wheat leaves. *Can. J. Bot.* **43**, 747-755.

Shaw, M., and Manocha, M. S. (1965b). The physiology of host-parasite relations. XV. Fine structure in rust-infected wheat leaves. *Can. J. Bot.* **43**, 1285-1292.

Simons, M. D. (1975). Effects of two systemic fungicides on ultrastructure of haustoria of the oat crown rust fungus. *Phytopathology* **65**, 388-392.

Skipp, R. A., Harder, D. E., and Samborski, D. J. (1974). Electron microscopy studies on infection of resistant (Sr6 gene) and susceptible near-isogenic wheat lines by *Puccinia graminis* f. sp. *tritici. Can. J. Bot.* **52**, 2615-2620.

Smetana, K., and Busch, H. (1974). The nucleolus and nucleolar DNA. *In* "The Cell Nucleus." (H. Busch, ed.), Vol. 1, pp. 73-147. Academic Press, New York.

Smith, L. D. (1975). Molecular events during oocyte maturation. *In* "The Biochemistry of Animal Development, Vol. III, Molecular Aspects of Animal Development" (R. Weber, ed.), pp. 1-46. Academic Press, New York.

Smith, J. E., Gull, K., Anderson, J. G., and Deans, S. G. (1976). Organelle changes during fungal spore germination. *In* "The Fungal Spore, Form and Function" (D. J. Weber and W. M. Hess, eds.), pp. 301-354. John Wiley and Sons, New York.

Stakman, E. C. (1914). A study in cereal rusts: physiological races. *Minn. Agric. Exp. Stat. Bull.* **138**.

Stanbridge, B., and Gay, J. L. (1969). An electron microscope examination of the surfaces of the uredospores of four races of *Puccinia striiformis. Trans. Br. Mycol. Soc.* **53**, 149-153.

Stoekenius, W. (1959). An electron microscope study of myelin figures. *J. Biophys. Biochem. Cytol.* **5**, 491-500.

Strobel, G. A. (1965). Biochemical and cytological processes associated with hydration of uredospores of *Puccinia striiformis. Phytopathology* **55**, 1219-1222.

Sussman, A. S., Lowry, R. J., Durkee, T. L., and Maheshwari, R. (1969). Ultrastructural studies of cold-dormant and germinating uredospores of *Puccinia graminis* var. *tritici. Can. J. Bot.* **47**, 2073-2078.

Takahashi, H., and Furuta, T. (1973). Observation on surface structure and behavior of *Puccinia coronata* Corda var. *avenae* Fraser et Leding. on the host plant by scanning electron microscopy. *Rept. Tottori Mycol. Inst.* **10**, 253-272.

Takahashi, H., and Furuta, T. (1975a). Scanning electron microscopic observations on uredospore development of *Puccinia graminis tritici. The Kanto-Tosan Plant Protection Soc.* **22**, 26-27. (In Japanese).

Takahashi, H., and Furuta, T. (1975b). Observations on teliospore germination and sporidium formation of *Gymnosporangium haraeanum* P. et H. Sydow by a scanning electron microscope. *Ann. Phytopathol. Soc. Jpn.* **41**, 69-72.

Taubenhaus, J. J. (1911). A contribution to our knowledge of the morphology and life history of *Puccinia malvacearum* Mont. *Phytopathology* **1**, 55-62.

Taylor, M. W. (1921). Internal aecia of *Puccinia albiperidia* Arthur. *Phytopathology* **11**, 343-344.

Thatcher, F. S. (1939). Osmotic and permeability relations in the nutrition of fungal parasites. *Am. J. Bot.* **26**, 449-458.

Thatcher, F. S. (1943). Cellular changes in relation to rust resistance. *Can. J. Res. Sec. C* **21**, 151-172.

Thomas, P. L., and Isaac, P. K. (1967). The development of echinulation in uredospores of wheat stem rust. *Can. J. Bot.* **45**, 287-289.

Trelease, R. N., Becker, W. M., Griber, P. J. and Newcomb, E. H. (1971). Microbodies (glyoxysomes and peroxisomes) in cucumber cotyledons. *Plant Physiol.* **48**, 461-475.

Trocha, P., and Daly, J. M. (1974). Cell walls of germinating uredospores. II. Carbohydrate polymers. *Plant Physiol.* **53**, 527-532.

Trocha, P., Daly, J. M. and Langenbach, R. J. (1974). Cell walls of germinating uredospores. I. Amino acid and carbohydrate constituents. *Plant Physiol.* **53**, 519-526.

True, P. R. (1938). Gall development on *Pinus sylvestris* attacked by the woodgate *Peridermium*, and morphology of the parasite. *Phytopathology* **28**, 24-49.

Tschen, J., and Fuchs, W. H. (1968). Endogene Aktivität der Enzyme in rostinfizierten Bohnenprimärblättern. *Phytopathol. Z.* **63**, 187-192.

Ushiyama, R., and Hashioka, Y. (1973). Viruses associated with hymenomycetes. I. Filamentous virus-like particles in the cells of a fruit body of shiitake, *Lentinus edodes* (Berk.) Sing. *Rept. Tottori Mycol. Inst.* **10**, 707-805.

Valkoun, J., and Bartos, P. (1974). Somatic chromosome number in *Puccinia recondita*. *Trans. Br. Mycol. Soc.* **63**, 187-189.

Vigil, E. L. (1973). Structure and function of plant microbodies. *Sub-Cell. Biochem.* **2**, 237-285.

Van Dyke, C. G., and Hooker, A. L. (1969). Ultrastructure of host and parasite in interactions of *Zea mays* with *Puccinia sorghi*. *Phytopathology* **59**, 1934-1946.

Walkinshaw, C. H., Hyde, J. H. and Van Zandt, J. (1967). Fine structure of queiscent and germinating aeciospores of *Cronartium fusiforme*. *J. Bacteriol.* **94**, 245-254.

Walles, B. (1974). Ultrastructure of the rust fungus *Peridermium pini* (Pers.) Lev. *Stud. For. Suec.* **122**, 1-30.

Waterhouse, W. L. (1921). Studies in the physiology of parasitism. VII. Infection of *Berberis vulgaris* by sporidia of *Puccinia graminis*. *Ann. Bot.* **35**, 557-564.

Welch, B. L., and Martin, N. E. (1973). Scanning electron microscopy of *Pinus monticola* bark infected with *Cronartium ribicola*. *Phytopathology* **63**, 1420-1421.

Welch, B. L., and Martin, N. E. (1974). Invasion mechanisms of *Cronartium ribicola* in *Pinus monticola* bark. *Phytopathology* **64**, 1541-1546.

Welch, B. L., and Martin, N. E. (1975). Light and transmission electron microscopy of the white pine blister rust canker. *Phytopathology* **65**, 681-685.

Wells, K. (1965). Ultrastructural features of developing and mature basidia and basidiospores of *Schizophyllum commune*. *Mycologia* **57**, 236-261.

Whitney, H. S., Shaw, M., and Naylor, J. M. (1962). The physiology of host-parasite relations. XII. A cytophotometric study of the distribution of DNA and RNA in rust infected leaves. *Can. J. Bot.* **40**, 1533-1544.

Williams, P. G. (1971). A new perspective of the axenic culture of *Puccinia graminis* f. sp. *tritici* from uredospores. *Phytopathology* **61**, 994-1002.

Williams, P. G., and Ledingham, G. A. (1964). Fine structure of wheat stem rust uredospores. *Can. J. Bot.* **42**, 1503-1508.

Willison, J. H. M., and Brown, R. M., Jr. (1977). An examination of the developing cotton fibre: wall and plasmalemma. *Protoplasma* **92**, 21-41.

Wilson, M., and Henderson, D. M. (1966). "British Rust Fungi." Cambridge University Press, London.

Wolf, F. A. (1913). Internal aecia. *Mycologia* **5**, 303-304.

Wood, R. K. S. (1973). Specificity in plant diseases. *In* "Fungal Pathogenicity and the Plant's Response" (R. J. W. Byrde and C. V. Cutting, eds.), pp. 1-16. Academic Press, New York.

Wright, R. G., Lennard, J. H., and Denham, D. (1978). Mitosis in *Puccinia striiformis*. II. Electron microscopy. *Trans. Br. Mycol. Soc.* **70**, 229-237.

Wynn, W. K. (1976). Appressorium formation over stomates by the bean rust fungus: Response to a surface contact stimulus. *Phytopathology* **66**, 136-146.

Wynn, W. K., and Gajdusek, C. (1968). Metabolism of glucomannan-protein during germination of bean rust spores. *Contrib. Boyce Thompson Inst.* **24**, 123-138.

Wynn, W. K., and Wilmot, V. A. (1977). Growth of germ tubes of the wheat stem rust fungus on leaves and leaf replicas. *Annu. Proc. Am. Phytopathol. Soc.* **4**, 129 (Abstr.).

Yaniv, Z., and Staples, R. C. (1969). Transfer activity of ribosomes from germinating uredospores. *Contrib. Boyce Thompson Inst.* **24**, 157-163.

Yarwood, C. E., and Hecht-Poinar, E. (1973). Viruses from rusts and mildews. *Phytopathology* **63**, 1111-1115.

Ziller, W. G. (1961). Pear rust (*Gymnosporangium fuscum*) in North America. *Plant Dis. Rept.* **45**, 90-94.

Zimmer, D. E. (1965). Rust infection and histological response of susceptible and resistant safflower. *Phytopathology* **55**, 296-301.

Zimmer, D. E. (1970). Fine structure of *Puccinia carthami* and the ultrastructural nature of exclusionary seedling-rust resistance of safflower. *Phytopathology* **60**, 1157-1163.

Supplementary
Bibliography *

Bennell, A. P., and Henderson, D. M. (1978). Urediniospore and teliospore development in *Tranzschelia* (Uredinales). *Trans. Br. Mycol. Soc.* **71,** 271-278.

Brown, M. F., and Brotzman, H. G. (1979). Phytopathogenic fungi: A scanning electron stereoscopic survey. University of Missouri (in press).

Codron, D. (1978). Etude ultrastructurale de la formation des eciospores chez le *Puccinia urticae-caricis* Kleb. *Rev. Mycol.* **42,** 77-96.

Guggenheim, R., and Harr, J. (1978). Contributions to the biology of *Hemileia vastatrix* II. SEM-investigations on sporulations of *Hemileia vastatrix* on leaf surfaces of *Coffea arabica. Phytopathol. Z.* **92,** 97-101.

Harr, J., and Guggenheim, R. (1978). Contributions to the biology of *Hemileia vastatrix* I. SEM-investigations on germination and infection of *Hemileia vastatrix* on leaf surfaces of *Coffea arabica.* Phytopathol. Z. **92,** 70-75.

Jonsson, A., Holmvall, M., and Walles, B. (1978). Ultrastructural studies of resistance mechanisms in *Pinus sylvestris* L. against *Melampsora pinitorqua* (Braun) Rostr. (pine twisting curl). *Stud. For. Suec.* **145,** 28 pp.

Peterson, R. S., and Oehrens, E. (1978). *Mikronegeria alba* (Uredinales). *Mycologia* **70,** 321-331.

Shaw, C. G., III (1976). Rust on *Phyllocladus trichomanoides*—the first recorded on a member of the Podocarpaceae. *Trans. Br. Mycol. Soc.* **67,** 506-509.

Wright, R. G., Lennard, J. H., and Denham, D. (1978). Mitosis in *Puccinia striiformis* II. Electron microscopy. *Trans. Br. Mycol. Soc.* **70,** 229-237.

*Subsequent to the submission of the manuscript, several significant publications dealing with rust ultrastructure have appeared and another, yet in press, has been brought to our attention. Another was inadvertently overlooked (Shaw, 1976). Significantly, peltate ornamentation of aeciospores, not described elsewhere ultrastructurally, was reported.

Taxonomic Index*

* Numbers in italic type denote pages with figures.

Subject Index

A

Aecium 20–31, 89, 91, 121
  intercalary cells, within 29, 31, 34, 35, 39, 89
  peridium of 21, 23–28, 30, 31, 41, 89
  types of
    aecidioid 21, 23–25, 44
    caeomoid 21–23, 89
    peridermioid 25, 26, 28, 29
    roestelioid 23, 25–27, 44, 89
    uredinoid 27, 91
Aecioid teliospore 3, 74, 75, 87
Aecioid urediospore 33, 45, 55
Aeciospore 1, 20–45, 88, 89, 127, 195
  cell wall of 37, 39, 41, 58, 88
  discharge from aecium 23, 25, 43–45
  formation of 27, 29, 31–35, 88
  germ pores of 39–42,
  germination and penetration of host 41, 89
  ornamentation 31, 33–39
  uredinoid 27, 35
Aeciosporophore 27, 29, 32–35
Amphispore 47
Apical vesicles 100, 105, 112, 123, 127, 149
  in germ tubes 100, 105, 127
  in infection structures 105
  in vegetative hyphae 123
Apiculus 4
Appressorium 67, 69, 93–100, 103, 104, 107, 114, 155, 239
  development 100, 114
  induction 94–98

Autoecious rust 1, 2
Autoradiographic studies
  of compatible interactions 191
  of incompatible interactions 215
Axenically cultured rust 2, 98, 105, 118, 123, 125, 126, 199–201, 247

B

Basidium, *see* Metabasidium
Basidiospore 1–5, 21, 85–90
  cell wall 4, 88, 89, 103
  germination 4, 5, 89,
  host penetration by germlings of 5, 89, 164, 166
  ornamentation 4, 88,
  structure 3, 4, 88, 89
Buffer cell, of telia 53

C

Caeoma 21–23, 89
Callose 186, 189, 217, 219, 225
Catalase 109, 126
Caviculum 87
Cell membrane, *see* Plasmalemma
Cell wall
  fungal
    of aecial peridium 44
    of aeciospore 37, 39, 41, 58, 88
    of aeciosporophore 29, 34, 35